PRAISE FO

SLIME

A *NEW YORK TIMES* NEW AND NOTEWORTHY TITLE

A *SCIENCE FRIDAY* BEST BOOK OF THE SUMMER

"Kassinger gives these Plain Janes their time in the sun . . . Great reading."
—Ira Flatow, NPR's *Science Friday*

"Algae are among the earth's oldest life-forms, pervasive in everything from pond scum to crude oil. Kassinger explains their history and biology, and makes a persuasive case for their future importance."
—*New York Times Book Review*

"No organisms are more important to life as we know it than algae. In *Slime,* Ruth Kassinger gives this underappreciated group its due. The result is engaging, occasionally icky, and deeply informative."
—Elizabeth Kolbert, *New York Times* best-selling author of the Pulitzer Prize winner *The Sixth Extinction*

"Kassinger has found in algae an undervalued topic truly worthy of closer attention . . . In the end, Kassinger has us rooting for pond scum—it might just save us yet."
—*Science News*

"In spite of having studied algae for more than thirty years, I learnt much from *Slime* . . . Compelling . . . There is something for everyone, from committed phycologists to people who hitherto (but hopefully no longer) regarded algae as an inconvenience or worse. Blanket weed may never seem the same again."
—*Nature*

"*Slime* illustrates the important role algae have played in the world . . . Overall, *Slime* gives a distinct view into these underappreciated organisms and demonstrates our intertwined history with algae. Hopefully, it will help readers see algae in a different light."
—*Science*

"Fascinating and relevant . . . As Kassinger finds unique nuggets within algae's backstory and possible future, she unravels amazing, microscopic details of this vital resource . . . Where it gets really interesting is her detailed explanation of the large role algae played in the complicated, multistep process of human evolution, supplementing our ancestors' diets with iodine and the omega-3 oil DHA (docosahexaenoic acid), both essential ingredients for developing larger brains. And it has continued to serve as a nutritious food source for many cultures ever since . . . Kassinger has penned a wondrous story of this multifaceted, often misunderstood microorganism whose existence is vital to our own."

—*BookPage,* starred review

"We often look for big solutions, but the reality is that the smallest things often offer hope. This globetrotting book showcases the 'algae innovators' (the phrase of the month) exploring what we can learn from these often-ignored plants." —*EcoWatch*

"A book full of delights and surprises. Algae are the hidden rulers of our world, giving us oxygen, food, and energy. This is a beautiful evocation of the many ways that our past and future are entangled in their emerald strands."

—David George Haskell, author of *The Songs of Trees*
and Pulitzer Prize finalist *The Forest Unseen*

"*Slime* is a revelation! Algae have the power to cool the planet, replace plastics, fuel vehicles, and feed the world. This visionary book belongs in the hands of every policy maker, business leader, and engaged citizen looking for answers to our most pressing problems. It also happens to be a delightful read in the tradition of Susan Orlean, Mary Roach, and Michael Pollan. Ruth Kassinger turns a reporter's eye to the natural world and finds an epic narrative there, populated by dedicated scientists, intrepid chefs, and starry-eyed visionaries."

—Amy Stewart, *New York Times* best-selling author of
The Drunken Botanist and the Kopp Sisters novels

"Deep and enlightening . . . Readers will learn more about algae than they ever imagined (and relish every minute of it). Comparisons to Mary Roach and Susan Orlean are well-deserved, and Kassinger's erudite and wide-ranging approach should entice readers with a wide range of interests, from food to fashion, bioengineering, marine biology, farming, and general fascination with the wonders of nature. Gardeners will welcome Kassinger's latest, and everyone else will feel lucky to discover this winsome writer."

—*Booklist,* starred review

"A fun and fascinating deep dive into the natural history, current uses, and vast potential of algae . . . Accessible and enthralling . . . Kassinger delivers the powerful and optimistic message that slime just may be our savior . . . Thorough but not dense, informative but never boring—a delight from start to finish."

—*Kirkus Reviews,* starred review

"Ruth Kassinger is a witty and affable guide throughout this globe-trotting celebration of an overlooked life form. Reading *Slime* will convince you that algae deserve our respect, and even our gratitude—they are ancient, fascinating, and essential to life as we know it."

—Thor Hanson, author of *Buzz, The Triumph of Seeds,* and *Feathers*

"In chirpy prose chock-full of homespun metaphors . . . Kassinger turns an obscure subject into delightful reading . . . Even readers who never expected to enjoy a book about slime will find this an informative and charming primer to 'the world's most powerful engines.'"

—*Publishers Weekly*

BOOKS BY RUTH KASSINGER

Paradise Under Glass

A Garden of Marvels

How Algae
Created Us,
Plague Us,
and
Just Might
Save Us

SLIME

RUTH
KASSINGER

Mariner Books
Houghton Mifflin Harcourt
Boston New York

First Mariner Books edition 2020

hmhbooks.com

Library of Congress Cataloging-in-Publication Data
Names: Kassinger, Ruth, author.
Title: Slime : how algae created us, plague us, and just might save us / Ruth Kassinger.
Description: Boston : Houghton Mifflin Harcourt, 2019. |
Includes bibliographical references and index.
Identifiers: LCCN 2018043835 (print) | LCCN 2018044781 (ebook) |
ISBN 9780544433151 (ebook) | ISBN 9780544432932 (hardcover) |
ISBN 9780358299561 (trade paper) | Subjects: LCSH: Algae.
Classification: LCC QK566 (ebook) | LCC QK566 .K37 2019 (print) |
DDC 579.8—dc23

LC record available at https://lccn.loc.gov/2018043835

Book design by Chrissy Kurpeski
Typeset in Chronicle Text

Printed in the United States of America
DOC 10 9 8 7 6 5 4 3 2 1

For Ted

Contents

SECTION III
PRACTICAL MATTERS

SECTION IV
ALGAE AND THE CHANGING CLIMATE

Introduction

Algae. When you read the word, what pops into your mind? A bright green ring of slime around an outdoor drain? Dark green fuzz obscuring the glass of a fish tank? Pea-soup scum blanketing a pond in midsummer?

Whatever unpleasant image it is, I understand. Before I wrote this book, if I'd heard *algae,* I'd have instantly recalled the greenish hems on the damp shower curtains in the girls' locker room at Pimlico Junior High. Seaweeds — these are algae, too — would have triggered another unpleasant memory. As a child, I learned to swim in a lake at summer camp, starting in the Guppies class and moving up to Minnows. The Minnows practiced a dozen yards from the shore, where the water was chest-high. Seaweeds grew in patches out there on the silty bottom, but I kept my eyes open underwater and usually managed to steer clear. Sometimes, though, if an earlier class had churned the shallows and clouded the water, I would accidentally blunder into them. Their caresses along my bare arms and legs would send me into a panic — the slippery vines seemed as if they might wrap around me and pull me under. I'd bolt to my feet, gasping, and flounder my way out.

After the age of eight (when I graduated to Sunfish and swam beyond the dock in deeper water), I rarely thought about algae. And then, quite suddenly, in December of 2008, I began to think about them a lot. That was when, in researching another book, I visited Valcent Products, a startup biofuel company operating in a greenhouse in the dusty outskirts of El Paso, Texas. Founder Glen Kertz, backed by millions of dollars in venture capital, was growing algae in dozens of vertical, clear plastic panels, each ten feet tall by four feet wide and a few inches thick. Inside the panels, a water-and-algae mixture flowed through serpentine channels. That day, with sunlight streaming down from overhead, the panels glowed a bright, almost otherworldly green. Kertz, a man built on the Daddy Warbucks model with candid blue eyes, explained that the color would deepen almost to black as the algae within doubled and redoubled. Once a day, half the contents of the panels was siphoned off and sent into a powerful centrifuge that separated the algae from the water, leaving an algal paste. The paste was then heated under high pressure to extract the oils inside, oils that would—when Valcent perfected its operations—be sold to a refinery to be processed into gasoline, diesel, or jet fuel. New water was added to the panels every day, so algae were perpetually on the job.

Success, Kertz asserted, was just around the corner; Valcent would soon be producing one hundred thousand gallons of fuel per acre. Like the many journalists writing awestruck articles about the company, I was ready to be convinced. After all, crude oil is made of ancient algae, compressed over millions of years below ground. Valcent would be doing in a short time what Earth has been doing, slowly, for ages. By burning algae fuel in our vehicles, we would be taking carbon dioxide *out* of the atmosphere rather than putting long-sequestered carbon *into* the atmosphere, as we do when we burn fossil fuel. Because 14 percent of global carbon dioxide emissions comes from burning transportation fuel, switching to algae oil would have a significant impact on climate change. Bonus points: growing algae requires no arable land and no fresh water, both valuable and increasingly scarce resources on our planet.

Unfortunately for his investors, Kertz wasn't much of a scientist or engineer, and shortly after my visit, Valcent went out of the oil business. (So much for those candid eyes.) Although Kertz's oil-per-acre figure was too high by a factor of at least ten, the science behind algae fuel is perfectly correct, and a host of other companies, mostly small startups but also Exxon, had begun developing the multiple technologies involved. I eagerly followed the progress of these entrepreneurs, trying to figure out what was really possible. As I did, I found myself increasingly engaged—and ultimately engrossed — by the little green cells at the heart of these enterprises. Biofuels, as it turns out, are only a small part of the promise of algae.

My fascination became this book. It is the story of my journey — by phone, plane, car, boat, drone, and scuba fins, across the US and around the globe from Canada to Wales to South Korea — to understand how algae, the most powerful organisms on the planet, influence our lives, for better and for worse, and what role they will play in our future. I start deep in Earth's history and travel to the bleeding edge of modern biotechnology. Along the way, I meet scientists and entrepreneurs who have been harnessing these tiny dynamos to improve our health, nourish our ever-growing human population, and clean up the mess we are making of our planet.

Algae are Earth's authentic alchemists. Using sunlight for power, they take the dross of carbon dioxide and, with water and a scintilla of minerals, turn it into organic matter, the stuff of life. Even better, as they work their combinatorial magic, they burp oxygen. Take a breath: At least 50 percent of the oxygen you inhale is made by algae. What is waste to them is priceless to all respiring animals. Without algae, we would gasp for air.

There is no shortage of algae. The oceans are blanketed in a dense but invisible six-hundred-foot-thick layer of them. There are more algae in the oceans than there are stars in all the galaxies in the universe. Swallow a single drop of seawater, and you could easily down several thousand of these unseen beings. They are the essential food of the microscopic grazing animals at the bottom of the marine food chain. If all algae died tomorrow, then all familiar aquatic life — from tiny krill to whales — would quickly starve.

In fact, if algae hadn't evolved more than 3 billion years ago and oxygenated the atmosphere, multicellular creatures would never have graced the oceans. It was a species of green algae that, 500 million years ago, acclimated to life on land and evolved into all of Earth's plants. Without plants to eat, the first marine animals that wriggled out of the water 360 million years ago would never have survived or continued to evolve and diversify into all the land-living creatures we know today, including us. If, several million years ago, our ancestor hominins hadn't had access to fish and other algae-eating aquatic life — and thus to certain key nutrients — we would never have evolved our outsize brains. Without algae, we couldn't know that all of life depends on algae.

Algae's influence extends long after their death. Their microscopic, carbonaceous corpses — those that don't become food for aquatic animals and bacteria — drift down through the ocean like a steady snowfall. On the ocean floor, they silently accumulate, their carbon remains sequestered for eons. By transferring carbon dioxide from the atmosphere to long-term storage, algae help keep our planet from becoming an unbearable hothouse. About fifty million years ago, when the Arctic was last ice-free year-round, a million-year burst of algae growth cooled the atmosphere to help create the icy conditions we see today.

Phycologists — scientists who study algae — have identified some 72,000 species, but there may be ten times as many yet to be named. Today, algae are in every earthly habitat, unseen under ice in Antarctica, blooming pink on the snows of the Sierra Nevada, in desert sand, inside rocks, on trees, and in the fur of three-toed sloths (who eat them). They live symbiotically inside corals, which cannot survive without their partners. Coral reefs are home to 25 percent of the world's fish species and provide economic support to hundreds of millions of people. Now they're dying at a shocking, heartbreaking rate because warming oceans disrupt the critical relationship between algae and their hosts.

So, what exactly are algae? There is no exact answer. *Algae* is not a precise term, not a taxonomic category like the kingdom Animalia or the genus *Homo* or the species *Homo sapiens.* Algae (and the singular alga) is a catchall term, a name for a group of diverse organisms. There are three types, which we'll meet in order of their evolution. The smallest, most ancient of them are the single-celled, internally simple blue-green algae, now generally known as cyanobacteria (or, more familiarly, cyanos). Next up are the invisible, single-celled but internally complex microalgae. Together, cyanobacteria and microalgae are sometimes referred to as phytoplankton, from the ancient Greek, meaning "plant drifters." Finally —

and flavorfully — are the visible seaweeds, or macroalgae. Whether a cyanobacterium a tenth of the width of a human hair or a giant kelp that grows more than 150 feet tall, algae share certain characteristics. Most prominently, almost all of them photosynthesize, and the few species that don't, once did.

Algae are also defined by what they are not. They are not plants, even though plants also photosynthesize. Until the twentieth century, algae were considered members of the kingdom Plantae. An understandable inclusion: many seaweeds look like plants, and colonies of microalgae growing on damp soil can look like mosses. Nonetheless, algae are fundamentally different from plants. They inhabit a kind of prelapsarian world where, floating in water with little or no effort, they bask in the sun's energy and bathe in waterborne nutrients that easily pass through their cell walls into their cytoplasm. You will never find algae dressed in flowers, wafting scents, or sporting seeds and berries. Plants are the fancy-pants photosynthesizers of our world; algae are the plain Janes. But because algae have no petals or nectar, no pistils or stamens, no bark to keep them from drying out, and no wood to hold them upright, they channel far more of the sun's energy into multiplying themselves. And that means they are dozens of times more productive than plants at making the carbohydrates, proteins, vitamins, and oils — as well as accumulating the minerals — that we value.

You can get algae's nutritional benefits — especially their healthy omega-3 oils — directly by eating seaweeds. Every year, more than twenty-five million tons of seaweed, with a value of more than $6 billion, are harvested from gigantic watery farms in East Asia or plucked from the rocks off the New England and northern European coasts. Seaweeds make up about 10 percent of Japanese and Korean diets, and their sales are growing worldwide. In the US various kinds of dried seaweeds are found on the shelves of grocery stores from Costco to Whole Foods, and they are sold just as widely in Europe. It's easy to see why seaweeds are so popular: not only are many highly nourishing, many contain savory umami, one of the five basic tastes our tongues perceive. I've included some seaweed recipes in the appendix if you're inclined to expand your culinary horizons.

Most of the algae we eat are macroalgae, but a number of companies are tinkering with microalgae and cyanobacteria, figuring out how best to process and market them for human consumption. Later in the book, I'll drop in on the test kitchen of a promising enterprise in San Francisco and sample cookies, bread, and other foods that contain algae protein and oil instead of eggs and butter. Spoiler alert: the results are satisfying.

Even if you don't choose to eat algae directly, you reap their nutritional benefit every time you eat seafood. Because sea animals dined on algae, they accumulated algae's omega-3 oils, so you benefit secondhand. But today, half the fish we eat are grown using aquaculture, where they are increasingly fed with corn and soy. Could we feed fish with microalgae and maintain their nutritional profile? A company in Brazil grows algae in steel vats for just that purpose, and I'll tour its operation to see how they do it.

Algae are a hidden part of our lives. You can find them in the kitchen: In ice cream to prevent ice crystals from forming, in chocolate milk to keep the cocoa suspended, in salad dressings to keep the components mixed, and in many other foods. Your tap water may have been filtered at the water treatment plant with live algae that remove nitrogen and phosphorus or with fossil algae that strain out particulate matter. Your fruits and vegetables may have been grown in soil supplemented with algae. You can find algae in the bathroom, where they thicken your lotions, keep your hair conditioner emulsified, gel your toothpaste, and coat your daily tablets. And you can now wear algae on your feet: a Mississippi company I'll drop in on is making the soles of running shoes from pond scum.

As useful as algae can be, there can be too much of a good thing. In our era of global warming and unabated fertilizer runoff, rampant algae overgrowth is taking over more of our lakes and bays. Some of these algae "blooms" are merely unsightly, but others poison animals, including us. Florida has been particularly hard hit in recent years. In 2018, the governor declared a state of emergency in seven counties as millions of dead fish washed up along the Gulf Coast and hospital visits for respiratory illnesses caused by airborne toxins jumped 50 percent.

Algae needn't produce toxins to kill. Indirectly, they create aquatic "dead zones," areas with little dissolved oxygen where nothing can live. There are now more than four hundred major dead zones around the world, covering tens of thousands of square miles, and they're expanding every year.

Algae's ability to multiply like mad threatens lives and livelihoods, but can we somehow harness that prodigious productivity for the benefit of the environment? Like firefighters who create back burns to fight forest fires, a Florida company I'll visit is battling a plague of algae with more algae. As the world's vehicles, factories, and power plants continue to pour carbon dioxide into the atmosphere, there may be a role for algae in that cleanup as well. Algae are scarcer in the Southern Ocean than in the others; scientists are investigating how we might expand their numbers and thereby pull more carbon dioxide from the atmosphere and sequester it on the ocean floor.

The story of algae is like a woody vine deeply rooted in the past, reaching out in many directions in the present, and sending out new tendrils to find new purchase in the future. To bring order to my luxuriant subject, I have organized it into four sections. In the first, I trace the birth of algae and their conquest of Earth. Next, I explore the pleasures of eating seaweed and meet some of the people in the multibillion-dollar business of bringing it to our tables. In the third section I tell the stories of people who have discovered all kinds of other uses for algae, from glassmaking in the seventeenth century to plastics and fuel today. Finally, I investigate the power of algae to alter — for worse, but maybe also for better — our overheated atmosphere and polluted waters.

But before we begin a three-billion-year journey to tomorrow, let me start just a few years ago and right in my own front yard.

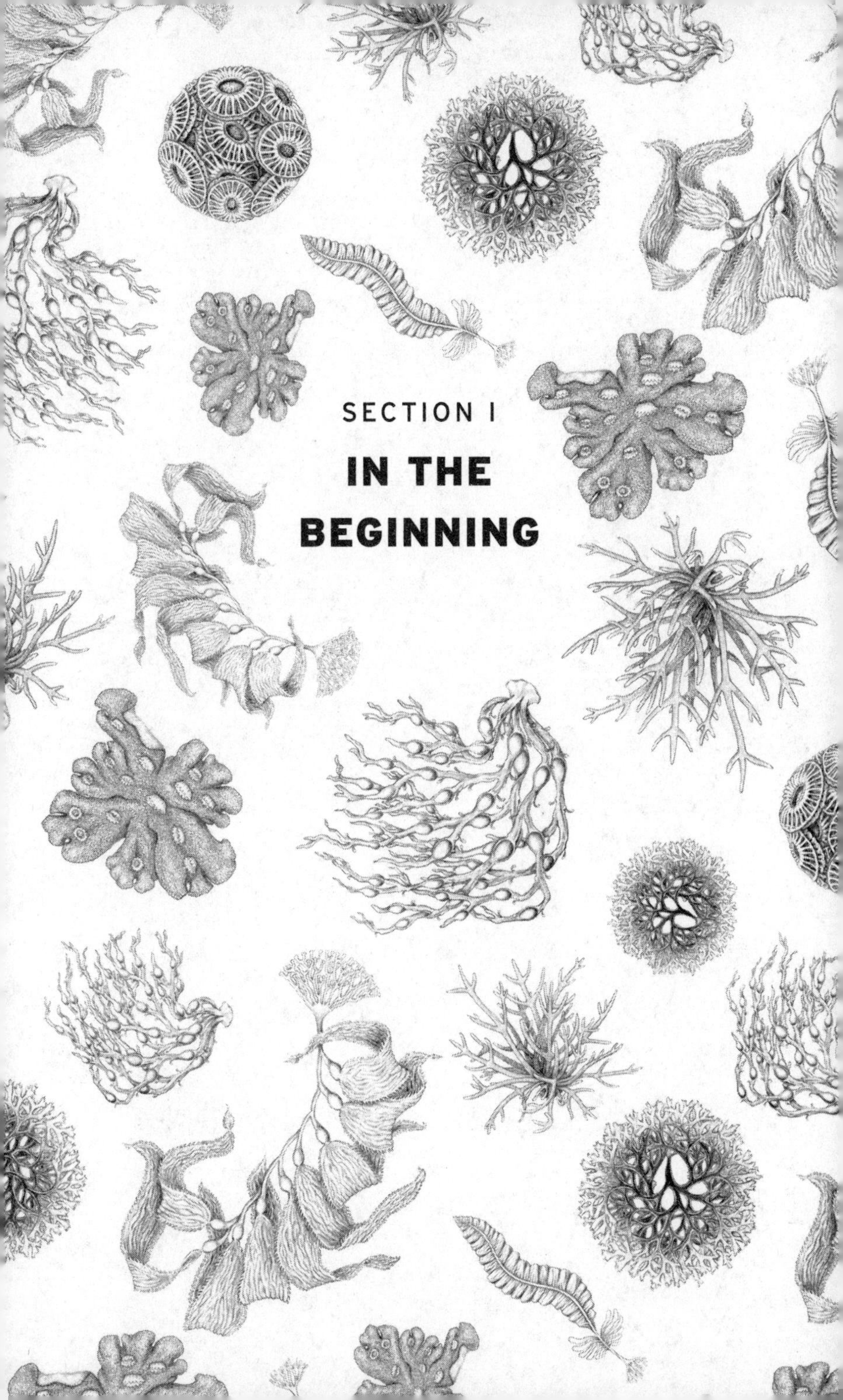

SECTION I

IN THE BEGINNING

1

Pond Life

A few years ago, I noticed an ad in a local magazine for a midcentury modern home designed by an architect I admired. Out of curiosity, I stopped by, even though my husband, Ted, and I weren't looking to move. I was immediately charmed by the glassy structure and especially by the pond—a square reflecting pool, really—that snugged right up to and across the front of the house. To get to the door, you had to walk across a pathway of slate pavers that appeared to hover just above the water's surface. Enchanting, I thought.

Enchanting indeed, for apparently it cast a benumbing spell on my rational brain. How else to explain that I didn't worry that the partial second story was tilted, seemingly intent on slinking off into the backyard, or that a soup pot sat on a bedroom floor, evidently to catch drips leaking through the roof? Why didn't I notice that the pond's concrete walls were crumbling, and that the basement wall that adjoined it was damp? Ted was less entranced with the architecture but was bewitched by the large and sunny yard pleading for the ministrations of a devoted gardener. Spellbound, we signed papers, wrote checks, and then spent the first eighteen months of ownership dealing with the house's structural problems. At last, Ted was able to turn his attention to the yard, and I focused on the pond.

Though neglected by humans, it was hardly abandoned by life. Reeds had grown up in the four basins sunk into its floor; dragonflies with iridescent wings darted among them. Spring peepers, brown tree frogs about the size of a quarter, serenaded us every evening with a high-pitched chorus. It was hard to spot the songsters; they clung to the mottled beige brick wall that borders one side of the pond and to the shrubs on the other. A few intrepid fellows climbed up the broad windows on either side of the front door, pasting themselves like decals on the glass. The peepers had relatives, big green bullfrogs with striped legs and bulging amber eyes who sat on the pavers and added bass notes to the afternoon and evening concerts. When I opened the door, the big guys marked my exit with a drumroll of plops as they hit the water.

All the frogs had come to the pond to mate, and evidence of their success – dozens of wide, gelatinous ribbons of eggs, each with a little black dot inside – soon appeared. Two weeks later, the dots hatched into thousands, maybe tens of thousands, of mobile commas. They emerged into an ideal incubator. With no fish around to feed on them, the roving punctuation survived in creepy numbers. But there was one benefit of this mass of tadpoles: We didn't have an algae problem. Young tadpoles gorge themselves on microalgae to fuel their rapid growth and metamorphosis. Our tadpoles were so numerous and hungry, the pond water was crystalline.

By September, the tadpoles had either died or turned into frogs or toads and moved on. I would soon miss them; in their absence the pond water underwent its own metamorphosis. First, I noticed a slight haze. The haze gradually deepened into a green murk. Then, one morning, I spotted crescents of scum collected in the pond's corners, and I recognized – with both irritation and some amusement, given my research – that I had grown my very own algal bloom. Under the summer sun, without any predators, algae had done what algae do so well: multiply. I could have battled the bloom with chemicals, but it was time to repair the concrete anyhow, so I pumped out all the water, along with all its microscopic inhabitants.

In the spring, the pond's only visitors were workmen. By July, the reconstruction was complete, and the rebuilt structure had an ultraviolet filter to zap suspended algae and six bubblers to keep the water gently rippling and oxygenated. My next step was to stock the pond. Even before we purchased the house, I'd pictured water lilies with yellow flower-bursts hovering above round, cloven leaves, and goldfish lazing about, vivid against a pitch-black pond floor. It was too late in the season to buy lilies, but a nearby aquarium store had plenty of goldfish, and I bought a mix of calico Shubunkins and piebald, orange-and-white Sarasas.

The sales clerk also urged me to buy floating ferns called azolla to provide shade for the fish in the midsummer heat and a hideout from any passing waterfowl interested in takeout sushi. That sounded like a purchase that a conscientious new parent of goldfish ought to make. But when the young man came back with a quart-size plastic bag of Lilliputian ferns, I thought he was kidding. The plants were pretty – in the bag, they looked like a bundle of green lace – but they were grossly inadequate for their job. It would be like me trying to shelter from a thunderstorm under a tropical drink parasol.

“No worries,” he assured me. “You’ll see. This stuff grows like crazy.”

Still dubious, I bought them, and as soon as I got home, I set the azolla and fish free. The fish flashed around in a school, evidently exulting in their escape from incarceration. The little ferns aggregated in a green disk the size of a dinner plate. By the end of the week, the plate had become a serving platter. Another week later, the platter was the size of a carpet sample. Within a month, I had a fuzzy green rug floating on the water, which was on its way to becoming wall-to-wall broadloom. By fall, I only saw a fish when one surfaced to nibble on the greenery. I had wanted a fishpond but had a fernpond instead.

I have since learned that, under ideal conditions, azolla can double in two to three days. The secret of the fern’s success lies inside the tiny leaves that compose its delicate, half-inch fronds. The leaves have an upper and a lower lobe. The lower one is translucent and shaped like a cup, so it acts like the hull of a boat and keeps the upper green lobe floating on the water’s surface. Each green lobe has a central cavity, and in that cavity live thousands of single-celled cyanobacteria, the simplest, most primitive type of algae. A particular species called *Anabaena azollae* lives exclusively in azolla. Not only are these cyanos photosynthetic, they also take nitrogen gas out of the atmosphere and transform it (that is, they “fix” it) into the nitrogen compounds that all creatures require to live. In exchange for safe harbor, the cyanos pass along much of their fixed nitrogen to their fern host. Terrestrial ferns use their roots to mine fixed nitrogen compounds from the soil, which means their growth is limited to what they can scavenge near their root tips. Azolla has no such constraints. Its cyanobacteria are in-home fertilizer factories, cranking out usable nitrogen all day.

Azolla's rocketing growth is not always appreciated; in certain countries and regions, ecologists label it as an invasive pest. In a sluggish canal or a placid reservoir, give azolla an inch, and it'll take an acre. It can block pump filters and cover a body of water so tightly that boaters find it impossible to make headway. But for all the trouble azolla can create, in East Asia it has ardent fans.

Takao Furuno is a man who loves azolla. For more than forty years, he has been growing rice on five acres of flat land near the town of Keisen on Kyushu, the southernmost island of Japan. Furuno is a wiry, warm, and earnest man, at least as far as I can tell by watching him in videos and reading his published work. He and his wife, who have raised five children, work together to grow and sell organic rice, vegetables, ducks, and duck eggs. Furuno is committed to the organic family farm, not only as a profitable economic endeavor, but as a spiritually satisfying and environmentally sound way of life.

Furuno was born in 1950 and grew up in the countryside of Kyushu. As a child, he played with his friends in the rice paddies and dug for loaches, a species of marketable, eel-like fish that burrowed into the wet mud when the paddies were drained. But by the time he was an adult and started his own farm, rice cultivation in Japan had changed completely. "We used to do it all by hand: weeding, harvesting, everything, but no longer." In the 1960s, farmers had turned to pesticides, herbicides, and synthetic fertilizers, and rice farming had become a monoculture. "Rice productivity rose and farmers' lives became much easier," Furuno says, "but there were no longer any fish in the paddies, and children no longer played there."

When Furuno took up rice farming in the early 1970s, he started out using chemicals, as all the farmers around him did. But after reading Rachel Carson's *Silent Spring,* he became concerned about the health of his young family and decided to take an organic approach. It was hard to make a financial go of it, though; weeds and insects ravaged his paddies. He was also working harder than his neighbors, rising before dawn and spending the entire day up to his shins in water, pulling weeds under the relentless summer sun. He tried all kinds of non-chemical methods of eliminating them: rotating between rice and vegetable crops, deep-water cultivation, double-plowing, releasing young carp, and using electric weeding machines. "But no matter what I did," he recalls, "the weeds defeated me."

Then, in 1988, a fellow organic farmer in a nearby town told him about an ancient Chinese method of duck farming. Domesticated ducks were allowed to graze freely in rice paddies, where they ignored the rice plants and fed themselves instead on weeds and insects. Furuno decided to give it a try. That summer, he released four-week-old ducklings into a newly planted paddy surrounded with netting. The results were spectacular: the weeds disappeared completely and the ducks prospered.

Furuno went all-in for his own version of the Chinese method, which he called integrated rice-duck farming. He bred his own ducks, a cross between small migratory wild drakes and heavier domesticated female ducks disinclined to fly. Although Chinese duck farmers traditionally sheltered their birds in sheds after dusk, he knew that ducks (unlike chickens) have good low-light vision and can forage on moonlit nights, so he kept them on the job around the clock. The extra hours his flock spent eating were helpful, but he kept losing ducks to the weasels, foxes, and feral and domestic dogs that prowled the area, as well as crows and other feathered foes who could snatch a duckling. "If the predators succeeded even once in invading the paddy field and tasted the delicious meat, they continued to try desperately to get the ducks." Furuno baited steel traps with raw eggs to capture weasels, and he hung netting from poles around the perimeter of the paddies, either draping it over the little levees to entangle intruders or embedding the bottom edges into the soil. He even tried sleeping by the paddies to drive off intruders. Finally, after three years of experiments, he developed a foolproof system that combined electric fencing, embedded netting, and fishing lines stretched in a zigzag pattern above the paddies.

Meanwhile, he perfected his rice-duck technique. He now transplants seedlings into his flooded paddies in early June and only introduces the fist-size ducklings three weeks later, when the young plants are large enough to stand up to the bump and bustle of the birds. Today he has about a hundred ducklings per acre that paddle about frenetically in the shallow water, unearthing weed seeds and stirring up silt that blocks the sunlight they need. Any weeds that do manage to poke their leaves up near the surface are nibbled relentlessly, if not uprooted altogether. The ducks also avidly pursue the various hoppers, cutworms, and borers that devour rice plants, either catching them outright or, in their boisterous, wing-flapping efforts to reach them, knocking them into the water to their death.

Word spread of Furuno's success in raising yields while eliminating the cost of pesticides and herbicides, and farmers across East Asia began to visit to learn his method. But he was not done tinkering. In 1993, he spoke with Professor Iwao Watanabe, an authority on azolla at the globally renowned International Rice Research Institute in the Philippines. Watanabe explained that for thousands of years, possibly as early as 6500 BC, Chinese rice farmers had purposefully cultivated azolla as a green manure. Their method was simple: Buy a few ferns in the spring from purveyors who learned to keep them alive through the winter, add them to the paddies — where they multiplied mightily — and, after draining the paddies in the fall, plow the azolla left behind into the soil. The azolla, rich in fixed nitrogen thanks to their cyanos, could double a paddy's productivity. Buddhist monks traveling throughout East Asia spread the practice, and eventually millions of acres of paddies were covered in a bright green carpet of the fern. As a fertilizer, it was effective and cheap.

In the 1960s, however, when chemistry came to farming, the new herbicides were as deadly to azolla as they were to weeds, and the ferns vanished from farms, their use forgotten. Professor Watanabe, having heard of Furuno's organic rice-duck method, thought the Japanese farmer might find growing azolla in paddies useful, not only as an organic fertilizer, but possibly as duck food.

Furuno tried it, and discovered that his ducks loved azolla. The ferns added enough additional calories to their diet that he could eliminate the supplemental feed he'd had to give them. (He still feeds his ducks a little broken, unmarketable rice, but merely to train them to come when he calls so he can easily gather them when he needs to.) In addition, he found that ducks feeding on azolla produce manure that is richer in fixed nitrogen. When the ducks' churning feet break up their own excreta, some of the nitrogen is scavenged by microalgae, which then flourish. The enriched microalgae become food for insect larvae, which in turn become more nutritious duck food. Any of the duck's nitrogen-rich poop that drifts down to the mud is taken up by the rice plants' roots, making for greener, healthier, higher-yielding rice plants. And the floating ferns don't compete with the plants for other nutrients because they draw them from the water, not the soil.

In 1996, Furuno added fish – the loaches he'd seen as a child – to the menagerie in his paddies. The fish also eat the microalgae and provide more manure for the rice, as well as revenue from sales. Furuno's rice-duck-azolla-fish method has now been adopted by more than 75,000 farmers in Southeast and South Asia, and that number is increasing. (It also earned him a PhD from Kyushu University.) Universities and research organizations across Asia confirm that farmers using his system have similar or higher yields than small farmers using conventional techniques. Moreover, they can sell their organic rice, as well as ducks and duck eggs, at a premium.

None of this would be possible without azolla, and azolla wouldn't exist without the cyanobacteria living inside them. It occurs to me that Furuno's rice paddy ecosystem is a microcosm of life on our planet: it all depends, and always has depended, on algae.

2

Something New Under the Sun

Imagine that you are a visitor on Earth 3.7 billion years ago, about 750 million years after the planet coalesced from cosmic dust. You are standing on a rocky volcanic island looking out over a greenish, iron-rich ocean that stretches to the horizon in all directions. There are no plants on your island; there is no soil for plants to grow in. Soil contains organic matter made of decomposed plant life, and no plants will grow on Earth for more than 3 billion years.

You will need to wear a breathing apparatus, as scuba divers do; as yet, there is no free oxygen on Earth. In fact, the atmosphere surrounding you is a deadly mix of carbon monoxide, carbon dioxide, methane, and possibly hydrogen and sulfur dioxide and nitrogen. But at least the temperature is pleasant: although the sun is only 70 percent as bright as it is today, the carbon dioxide and methane keep the surface of the planet warm. Earth is rotating twice as fast as it does now, so you may be disconcerted to see the sun rise and set every six hours. The moon is also ten times closer and looms ten times larger and brighter in the sky. Its proximity gives it great gravitational power, pulling tides hundreds of feet in and out. You should be able to see its cratered surface clearly, but perhaps not this day. Volcanoes are far more active around the globe than they are now, regularly lofting ash and droplets of sulfuric acid into the atmosphere. At dawn and dusk, the sky glows shades of yellow and tangerine.

The world's oceans have twice as much water as they do today, but the water is disappearing molecule by molecule. Because there is no oxygen in the atmosphere, there is no ozone layer. Without an ozone layer, ultraviolet radiation from the sun pours down on the planet unimpeded and blasts water into its components of hydrogen and oxygen. The lightweight hydrogen atoms go speeding off into space, and the oxygen immediately bonds with minerals in the water. Without an ozone layer, Earth is heading toward arid lifelessness, a fate that is befalling watery Venus, which is now bone-dry.

Nonetheless, at this time, there are oceans on our planet, oceans that harbor simple, single-celled bacteria and their single-celled cousins, the archaea. Like all living beings, these microscopic creatures need energy to function, as well as to construct more cell components when they are ready to divide to reproduce. Because their cell walls are rigid, eating their fellow creatures for energy is impossible. They can, however, pull compounds like hydrogen sulfide through their cell walls and into their interiors, where the compounds chemically react to release electrons. They use the electrons to create a short-term store of energy in a molecule called ATP (adenosine triphosphate). Then, using the energy of ATP and carbon dioxide dissolved in the water, the organisms build the organic compounds, including amino acids, proteins, lipids, and carbohydrates, that they need to live and reproduce.

There are still plenty of such “chemoautotrophs” around today, living in extreme environments like deep-sea hydrothermal vents or the hot, sulfur-rich springs in Yellowstone National Park. But sometime around the day of your visit (plus or minus a hundred million or even a couple hundred million years), a bacterium evolved that was truly something new under the sun. It floated near the surface of the ocean and appeared blue-green (cyan) thanks to its chlorophyll and other pigments. The pigments trapped photons, the particles of solar energy, which the cyanobacterium used to split water into hydrogen and oxygen, generate electrons, and make ATP. Then, just as the chemoautotrophs did, it used ATP to construct organic compounds. The process, of course, is photosynthesis, and because the cyanobacterium burps oxygen as waste, it is called oxygenic photosynthesis. The process is so complex that scientists are still uncovering the details of its mechanisms.

Cyanobacteria were ideally equipped to prosper. Whereas archaea and all other bacteria simply floated around, hoping to bump into their favorite chemical food, these new kids on the block, rather than splitting molecules they might happen upon *in* the water, split the ubiquitous water molecules themselves. Eating whenever the sun shone, the cyanos multiplied extravagantly. And the oxygen they bubbled out will eventually (as in a couple billion years) float into the atmosphere, create a protective ozone layer, and save our blue planet from a dusty death.*

As if that weren't enough, some species of cyanobacteria had more tricks up their blue-green sleeves. Life on Earth requires nitrogen; nitrogen is in DNA, ATP, and proteins, among other compounds that living things absolutely require. Earth's atmosphere has long been rich in the gas, but nitrogen atoms bind tightly to each other in molecular pairs, and life can't use N_2. Lightning strikes, blazing with 100 million volts or more of energy, can split N_2 in two and combine the individual atoms with hydrogen and/or oxygen to make the fixed nitrogen compounds of ammonia, ammonium, and nitrates. The trouble is, while lightning strikes are impressive, they're not particularly munificent in creating these compounds. If life had relied on lightning alone, it would never have gotten off — or onto — the ground.

Enter cyanobacteria. They were able to accomplish exactly the same thing as lightning, albeit on a microscopic scale. Cyanos became the principal nitrogen fixers (diazotrophs) on the planet. Fortunately, they were happy to share what they made; they leaked about 50 percent of their fixed nitrogen trove into the water, where other bacteria and archaea could take it up. Without diazotrophic cyanobacteria, only the simplest marine life forms would exist — and not too many of them at that. There simply wouldn't be enough fixed nitrogen to go around.

Cyanos get extra kudos because fixed nitrogen wasn't — and

*The very first photosynthesizers did not produce oxygen. Instead, they captured near-infrared light with a purplish pigment, stole electrons from sulfur compounds in the water, and pooped microscopic grains of pure sulfur as waste. They never prospered the way their cyano cousins did, but their descendants carry on in anaerobic waters today.

isn't—easy for them to make. First, they had to manufacture an enzyme called nitrogenase, which contains iron and molybdenum and catalyzes the reaction that fixes nitrogen. Then, they had to circumvent a problem of their own making, oxygen. Here's the rub: An oxygen atom has six electrons in its outer shell and is always quick to grab two more to achieve its full complement of eight. Early ocean waters were full of dissolved iron, which has two electrons in its outer shell. You can see what happened: a cyanobacterium's waste oxygen immediately grabbed the iron in a chemical embrace, leaving none for the cyanobacterium to employ to make nitrogenase.

Cyanos had to get creative. Some stopped photosynthesizing—and therefore emitting oxygen—while they fixed nitrogen. Others fixed nitrogen only at night while they weren't photosynthesizing (thereby making hay while the sun doesn't shine). Some joined forces with others of the same species to form microscopic filaments, like a string of beads. Every tenth or so bead stopped photosynthesizing and produced extra cell walls that made it impervious to oxygen. These beads ("heterocysts") devoted themselves to fixing nitrogen, and they survived by sharing their output with their next-door neighbors in exchange for sugars. Today, diazotrophic cyanobacteria still use one of these solutions.

But nitrogen was not the only problem cyanos had to solve to make it in the world. They faced another dilemma: They needed to gather sunlight near the ocean's surface without allowing ultraviolet radiation to fry their DNA. In response, they evolved a slick exterior layer of polysaccharides (long chains of sugar molecules) called mucilage that acted as the world's first sunscreen—and gave them their trademark slick surface. Ultimately, all algae would come to dress themselves in slime.

All in all, cyanobacteria were superbly equipped to prosper. Most species can double every seven to twelve hours, which means a one-by-one-foot plot covered in cyanos would cover the floor of a small office in two days. A few species can double every two hours, which means they could multiply to cover more than six football fields in the same period. Either way, with time measured in hundreds of millions of years, the first cyanobacteria multiplied beyond all imagination. Along the way, they evolved into species with a multitude of sizes and shapes: balls and ovoids, rods and spirals and filaments. (The smallest and most prolific, the round *Prochlorococcus*, was only discovered in 1986; there can be as many as four hundred thousand of them in a single teaspoon of ocean water.) Floating near the surface and stuck together by their mucilage, they formed masses of floating green mats. The mats assimilated the microscopic flotsam and jetsam of the era, which included common minerals like calcium carbonate and magnesium carbonate, as well as dead microorganisms, and grew denser. All the while, living cyanobacteria glided — thanks again to their mucilage — toward the sunlit surface and kept multiplying.

Cyanos proliferated most abundantly in shallow waters where essential minerals like phosphorus and molybdenum eroded from the rocks. Year by year, half-millimeter by half-millimeter, the mats grew deeper, and all but the top surface gradually hardened into layered rock formations. Over the eons, these formations, called stromatolites, grew until they were hundreds of feet deep and miles long. Some formed domes the size of hills. Slice through an ancient stromatolite, and you see what look like countless layers of tissue paper, each a season's growth, in tones of taupe, sand, and ochre.*

The exponential growth of cyanobacteria continued unimpeded for about 1.5 billion years. Because they faced no predators, cyanobacteria kept reproducing, absorbing carbon dioxide, creating fixed nitrogen, and burping oxygen into the water. None of that oxygen entered the atmosphere, though. All of it latched onto dissolved iron molecules, creating iron oxide. In other words, the ocean rusted. Over the course of a billion years, cyanobacteria sent twenty times more oxygen than there is in the air today into the

---

*For the most part, stromatolites stopped forming hundreds of millions of years ago, when fish and other marine animals evolved and began grazing the living, topmost layer. But you can still see them in places like Shark Bay, Australia, where the water is too saline for fish or snails to live. At low tide, the stromatolites look like boulders sitting on the sand, with oxygen bubbles rising from their wet surface.

oceans, where it hooked up with iron and drifted down to the ocean floor in reddish particles. Ultimately, cyanos created about 85 billion tons of iron ore, an amount equal to 5 percent of the Earth's crust. Next time you open your steel car door or pick up a stainless steel fork, consider the cyanobacteria that helped make it.

One day, about 2.5 billion years ago, cyanobacteria's oxygen combined with all of the oxidizable metals in the ocean and the first bubbles of oxygen drifted into the air. Free oxygen! Surely the curtain would now go up on complex life, fish would soon be swimming the seas, and amphibians crawling onto shore. But not yet; not for millions more years. Because landmasses were rich in iron too, the first atmospheric oxygen went to rusting rock. Finally, with nothing left to bond with, the first molecules of oxygen slipped out of the water to float free into the atmosphere.

Paradoxically, what we think of as life-giving oxygen was actually catastrophic for living beings. Oxygen was toxic to bacteria and archaea that had evolved to live in anoxic water, so these species either died, mutated to use oxygen for respiration, or relocated to the ocean depths, where oxygen hadn't penetrated. But cyanobacteria had another powerful means to deal out destruction: lethal ice.

As oxygen molecules drifted into the atmosphere, some reacted with methane gas, transforming it into carbon dioxide and water. Both methane and carbon dioxide are greenhouse gases that trap solar heat in the atmosphere and prevent it from radiating back out to space, but methane has a far more powerful greenhouse effect. So, as methane became CO_2, more heat escaped to space. At the same time, cyanobacteria continued to soak up carbon dioxide through photosynthesis and send large amounts of it to the ocean floor in the form of dead cells. In sum, cyanobacteria were slowly removing two greenhouse gases from the biosphere, and therefore cooling the planet.

Accelerating the process, the supercontinent of the era, Kenorland, was breaking apart into smaller continents and exposing more rock surface to the atmosphere. More rock surface meant more weathering, a process that also draws carbon dioxide out of the atmosphere. As a result, Earth grew colder and ice slowly advanced from its poles toward the equator until, about 2.4 billion years ago, it had become a giant snowball. In some places, ice and snow lay half a mile deep.

Microscopic algae, innocuous in the singular and extraordinarily powerful in the unfathomable many, had first conquered and then killed a living planet.

3

Algae Get Complicated

The thing is, it's hard to keep a planet cold if it's got a hot heart. During the 300-million-year Huronian glaciation, while Earth's surface lay icebound, magma and hot gases continued to rise from the planet's molten mantle through volcanoes and rifts in the ocean floor. New methane and carbon dioxide entered the water and the atmosphere, gradually reversing the work of cyanobacteria. About 2.1 billion years ago, the atmosphere had trapped enough heat to defrost the planet. The cyanobacteria that had survived the big freeze – perhaps under thinner ice in equatorial regions or near the shores of active volcanoes – got back to work.

At this time, an entirely new life form evolved in the oceans. Previously, the only living creatures had been bacteria and archaea, both of which are prokaryotes: simple, single-celled creatures that have a rigid cell wall, no membrane-bounded nucleus and no organelles, and whose DNA is contained in a single circular chromosome. The newbie was a eukaryote, a far more sophisticated organism with a membrane-bounded nucleus that contained multiple linear chromosomes, as well as a host of other organelles. Chief among these little internal organs were mitochondria, which are oxygen-burning, internal combustion engines that convert food into energy. Mitochondria turbocharged the eukaryotes, enabling them to construct a hundred thousand times more proteins – and therefore more enzymes, hormones, and structures – than prokaryotes. It was as if the new organisms had the power of a sophisticated 3D printer whereas cyanos had a starter set of LEGOs. Among the important new structures eukaryotes built was a flexible internal framework called a cytoskeleton and whirling flagella that propel them in their hunt for food. And hunting is the appropriate word. Prokaryotes "eat" by transporting small molecules through narrow channels in their cell wall, but the more elastic eukaryotes, which are ten to twenty times larger than prokaryotes, can eat by engulfing entire organisms.

One day, roughly 1.6 billion years ago, one of these hefty, well-equipped eukaryotes bumped into a tasty cyanobacterium. The predator consumed its prey, as single-celled eukaryotes do, by first folding its wall around it. The fold became a pouch, which then closed and pinched off from the cell wall. The cyanobacterium found itself inside the predator, inside a vacuole—a bubble, of sorts—formed from a piece of the eukaryote's cell wall.

This should have been the end of the cyanobacterium. The eukaryote's digestive organelles—its lysosomes—should have latched onto the vacuole and broken its contents down completely. But on this one day, at this one unique moment, the victim somehow evaded destruction. The cyanobacterium took up residence inside its would-be predator. There it lived on, still intact inside its own membrane, as well as a second membrane, the "gift" of the eukaryote.

Now the roles changed. The predator became a landlord; the victim its tenant. The eukaryote sheltered the cyanobacterium; the cyanobacterium, using sunlight that passed through the eukaryote's cell wall, continued to make sugars. Some of those sugars it leaked into its host—rent payments, you could say. So cozy and comfortable was the cyanobacterium in its new home that it continued to reproduce. And when the eukaryote reproduced, its offspring inherited encapsulated cyanobacteria, like buying a house and having the rent-paying lodgers stay on. Over time, some of the cyanobacteria's genes transferred into the nuclei of their hosts. This was a survival imperative; DNA replication takes time and energy, and those individual eukaryotes that eliminated duplicative genes could reproduce faster and more efficiently, and thus become more numerous, which is to say, more successful.

Eventually, the cyanobacteria lost so many of their genes—about 90 percent—to their hosts that they could no longer function independently. They had become chloroplasts, the green, disk-shaped organelles that carry out photosynthesis. These eukaryotes could now feed themselves with the energy of the sun, making them photoautotrophs, and they ceased to be heterotrophs, organisms that feed themselves by eating other organisms. More precisely, they were eukaryotic, single-celled, oxygen-emitting photoautotrophs. In other words, they were now microalgae.

Microalgae evolved in other ways. They developed pyrenoids, bubble-like compartments in the cell that concentrate carbon dioxide around an enzyme called Rubisco, the enzyme at the heart of photosynthesis. In photoautotrophs, Rubisco catalyzes the chemical reaction that converts carbon dioxide into organic compounds. With pyrenoids, microalgae were cooking on a professional range instead of a hot plate.

They also split into two major lineages in this period: red algae (or rhodophytes) and green algae (or chlorophytes). The third major group, the brown algae (Phaeophyceae) – likely a fusion between a red and a green alga – would evolve much later. Don't be misled, though, by these color categories. Algae and seaweeds contain a variety of pigments. Not only are some pigments more dominant than others, their intensity can vary with changing environmental conditions. Red algae can look dark green, purplish, and even black; brown algae can appear green, red, yellow, or even blue. What you see is not necessarily what you've got.

Microalgae were far more sophisticated, powerful, and adaptable than cyanobacteria, and they seemed poised to take over the oceans. But poised they remained. Despite their advantages, they wouldn't flourish for a billion years. From about 1.8 billion years ago to 800 million years ago (an era known as the Boring Billion), the simpler, smaller cyanobacteria continued to rule the waves.

What held microalgae back? For one, they can't fix nitrogen because they can't manufacture nitrogenase. They never developed mechanisms for isolating the waste oxygen they produce, which meant they had to scour nitrogen, already fixed, from the water. Today, oceans, rivers, and lakes have higher levels of fixed nitrogen, courtesy of cyanobacteria, but during the Boring Billion, the compound was in short supply.*

Fixed nitrogen wasn't the only molecule holding microalgae back. They were also in dire need of phosphorus and other mineral compounds. Imagine two porous ceramic balls, one the size of a golf ball (a cyanobacterium) and the other the size of a large beachball (a microalga). Drop them into a tub of red dye. It will take longer for the beachball to absorb the dye and turn completely red than for the golf ball. In fact, if there's not enough dye, the larger ball will never become fully red. Substitute essential minerals for dye in this thought experiment and you see the problem — the large microalgae required more nutrients than the surrounding water could supply. To make matters worse, microalgae need greater amounts of nutrients per volume than cyanobacteria do, thanks to their sophisticated structures. Even today, the nutrient-poor areas of the ocean are dominated by cyanobacteria. So, although the Boring Billion was an era of almost perpetual summertime, the living wasn't easy for microalgae. Despite their advantages, they merely puttered along in shallow waters. The world was dominated by prokaryotes, and life made no great leap forward.

*No eukaryote has ever been able to fix nitrogen. While certain plants — legumes such as peanuts and soybeans — are said to fix nitrogen, diazotrophic bacteria living on the plants' roots actually do the job for them, in exchange for sugars the plants leak from their roots.

4

Land Ho, Going Once

Finally, life got more lively. How so? The Earth moved.

Earth's crust is broken up into a dozen rocky tectonic plates that move about on a softer underlying mantle. Some of the plates carry major landmasses; others are covered primarily by water. At the beginning of the Boring Billion, the continent-carrying plates got locked together, like cars jammed into an intersection when the traffic light has failed, and formed a single, massive continent. At the end of the period, about 800 million years ago, rifts opened up at the plates' junctures, and the continents slowly moved apart. As they did, more coastline was exposed to currents and tides. Erosion increased dramatically, spilling greater amounts of minerals into offshore waters. And, according to scientists Jochen Brocks and Amber Jarrett at the Australian National University, the retreat of massive glaciers at the end of a major glaciation (the Sturtian) scraped phosphorus and other nutrients from rocks, which then washed into the oceans. The bottom line: more minerals fed more cyanobacteria, and more cyanobacteria meant more fixed nitrogen in the water.

All these nutrients were microalgae's ticket to reproduce. Their numbers exploded, and they—along with cyanos—created vast, thick slicks of green scum along the new coasts.

At the same time, microalgae began to evolve, transforming themselves into the beautiful, exotic zoo of microscopic life forms that we see today. A group of round species, the coccolithophores, grabbed calcium carbonate from the water and turned it into armor of exquisitely sculpted shields and spikes that make the organisms look like carved ivory beads. (The skeletons of coccolithophores accumulated on the ocean floor for millions of years and are visible today, constituting much of the white cliffs of Dover.) Diatoms incorporated silicon into their cell walls, making crystalline shells that have upper and lower halves —or frustules—that fit just inside each other like a box and its lid. (When diatoms reproduce, their halves split apart and each one develops a new frustule mate.) Dinoflagellates evolved two flagella, hairlike structures that whirl with different rhythms and allow them to move forward and to make controlled turns. Some microalgae evolved toxins; others luminescent compounds, perhaps to confuse predators. One group evolved an ability to thrive in hot water (up to 56 degrees Celsius or 132 degrees Fahrenheit) and tolerate sulfuric acid, arsenic, and other heavy metals; another developed a flowing motion that allows individuals to ooze their way through sand and mud. Still others reproduced as single cells but joined together to form filaments, which likely de-

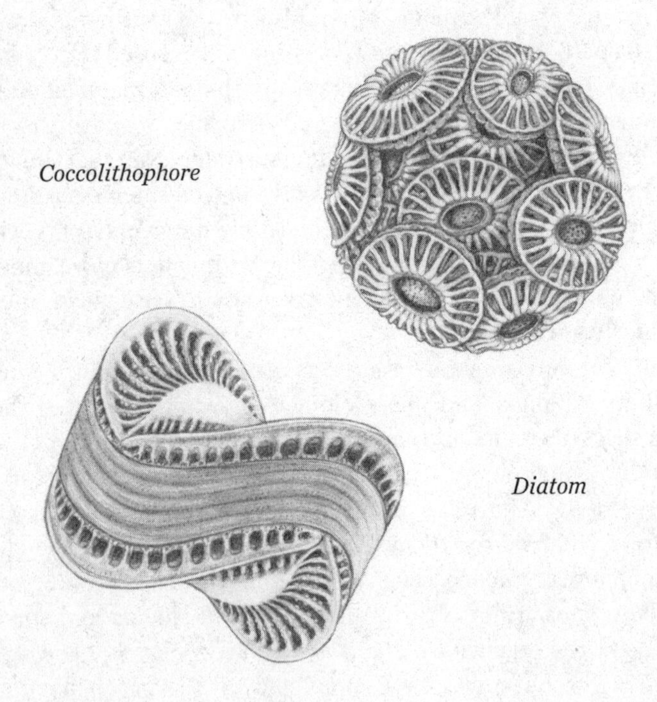

Coccolithophore

Diatom

terred predators who found them too large and difficult to consume.

Then, 650 million years ago, life took another big leap: some single-celled organisms became multicellular. (Multicellular organisms are those whose cells are permanently linked to one another, communicate among themselves, and perform different functions.) As multicellular creatures ourselves, the advantages of this body form seem obvious, but they aren't really. For billions of years, life existed only in single-celled forms, which were extraordinarily successful. The question of why and how multicellularity evolved is one of the most interesting questions in evolutionary biology.

Biologist Steven M. Stanley at Johns Hopkins University suggested in 1973 that it began with microalgae. He proposed that small microalgae in danger of being eaten by larger, single-celled heterotrophs somehow banded together to make themselves harder to be engulfed. The theory is appealing: everyone intuitively understands that walking in a group through dangerous territory is a better idea than going it alone. But how exactly would the transition have occurred?

In the late 1990s, Martin Boraas, professor at the University of Wisconsin, re-created a process that might have started the journey. He and his colleagues had long been growing *Chlorella vulgaris*, a common green microalgae species, as a sort of guinea pig for their lab experiments. Ordinarily, an individual *C. vulgaris* reproduces by growing larger while simultaneously dividing into at least two and as many as sixteen daughter cells. The daughters then separate one from another, tearing apart the cell wall of the mother as they do, and quickly grow to adult size. Generation after generation, the microalgae reproduce in this manner. But once in a long while, a mutation leads to an incomplete separation of the daughter cells, which instead remain loosely clustered together and linked by a ropy membrane that is the remnant of the mother's cell wall.

In his experiment, Boraas introduced a common single-celled predator, *Ochromonas vallescia*, into a microalgae culture. *O. vallescia* are twice the size of the microalgae, so they readily grazed them and, well fed, they reproduced with abandon. But once the predators had polished off most of their prey, their numbers declined sharply. Not surprisingly, the microalgae rebounded. These seesawing population explosions continued, but over time, the composition of the *C. vulgaris* community changed. More and more of the offspring emerged from the mother cell not as single cells, but in those loosely attached clumps — clumps that, not coincidentally, were too large for the *O. vallescia* to consume. By the twentieth reproductive cycle, most of the microalgae reproduced this way. Even when predator numbers were low, the cluster form of microalgae prevailed. Boraas and others hypothesize that this sort of defensive strategy was a first step in the evolution of multicellular algae — seaweeds.

Seaweeds, or macroalgae, are more than conglomerations of microalgae. For one, they have three distinct parts of their body, or thallus. Starting from the top, the blades of a seaweed are the organism's chief light-capturing device, much like a plant's leaves. (Some seaweeds' blades branch, forming a treelike shape; others are a single pennant.) Many seaweeds have a second body structure, the stipe, which is a flexible, hollow, stemlike part that keeps the seaweed upright. In large seaweeds, the stipe also channels sugars from the most photosynthetically active blades near the water's surface to the lower blades, which can be (as in the case of the giant kelp) more than 150 feet from sunlight. Most seaweeds also have a holdfast, which anchors the seaweed to sand, mud, or rock, either with a kind of glue or with rootlike projections. Unlike roots, however, a holdfast does not take up nutrients. Some seaweed species also have pneumatocysts, which are round or oblong gas-filled sacs that float the blades toward the surface, where photons are more plentiful. All the cells in the blades and stipes photosynthesize and take in minerals through their cell walls.

It's hard to precisely date the stages in algal evolution because these organisms, including seaweeds, generally don't fossilize. But in 2010, paleontologists were excavating a site near Lantian in central China that had once been a coastal zone. Peeling back thin lay-

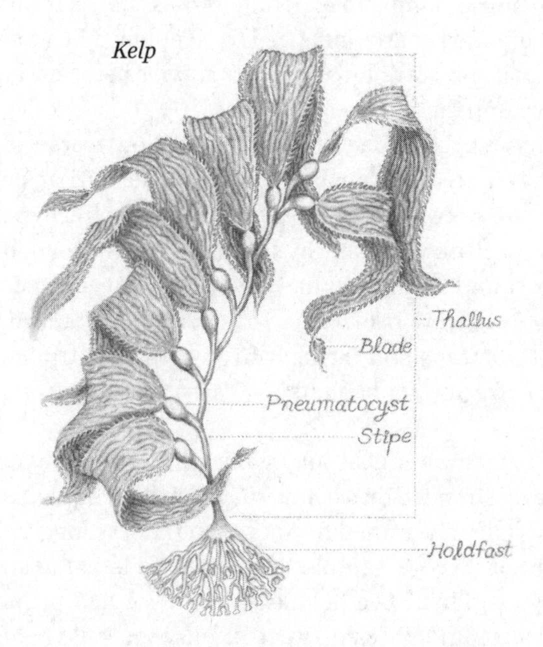

ers laid down roughly six hundred million years ago, they uncovered the carbon shadows of thousands of inch-high seaweeds that had a narrow stemlike stipe and delicate, branching blades that look, in carbon profile, like twigs. In fact, the Lantian seaweeds look so much like tiny plants, it's tempting to imagine that they simply crept onto land to become Earth's first plants. But seaweeds were too specialized and had traveled too far down their evolutionary path to adapt to such a radically different environment. It would be like fish evolving directly into mammals without having passed through amphibian and reptilian incarnations. In fact, plants evolved from a far simpler group of organisms: a group of single-celled green microalgae called charophytes.

The first step in the ultimate (but not inevitable) path to land was that some marine charophytes evolved to survive in fresh water. This was no small feat; oceans are dense with minerals that microalgae take up easily—so easily, in fact, that they developed electrochemical pumps to expel the excess. In fresh water, however, minerals are scarce and need to be pumped in, not out. This meant that if any saltwater alga was to survive in fresh water, it had to have a pump that worked in reverse.

Scientists suggest that, around 730 million years ago, certain charophyte individuals made just such an adaptation, perhaps in deltas where saltwater bays were infiltrated by freshwater rivers or, having been blown inland by storms, in freshwater ponds. In either case, some charophytes had a mutation that enabled their cellular pumps to work backwards. In fact, today certain charophyte species (the euryhalines) are able to quickly adapt from marine to brackish to freshwater habitats.

Imagine a freshwater charophyte microalga in a pond or stream, temporarily stranded on a damp, shaded bank. The alga could survive, especially if it rained from time to time, as long as it wasn't blown out of its covert into searing sun and lethal heat. Ancient charophytes didn't have holdfasts, but some had primitive rhizoids, delicate little threadlike extrusions that could lightly snag a surface. When an alga with rhizoids found itself stranded or blown onto land, its rhizoids helped it stay in place. Modern plants have inherited the genes that code for rhizoids; today, those ancient genes play a role in developing and maintaining roots.

Anchorage was only part of the value of rhizoids. Landlocked algae needed a way to lift water up from moist soil, against the force of gravity, and into the rest of their thallus. In rhizoids, they had just such a mechanism. But how could rhizoids lift water when they didn't have a hollow core, as roots do, to channel water upward? Try dipping the lower end of a paper towel into water, then hold it up and watch a liquid defy gravity. The fibers in the paper lie so close to one another that capillary action wicks water up. Charophytes with multiple rhizoids could have lifted water a short distance effortlessly.

These protoplants would have obtained some of their mineral nutrients from groundwater, but most plants need more than groundwater can supply. Today, 90 percent of plant species capture additional minerals thanks to certain rod-shaped soil fungi, known as mycorrhizae, that infiltrate the superfine hairs at the ends of root tips. A mycorrhiza, with one end in the soil and the other in the root hair, acts like a two-way shuttle: the soil-embedded end mines and passes along minerals and water to the root; the plant-embedded end passes along sugars to the fungus in return. While plants generally fight fungal infestations, they have genes that sense a protein uniquely present in mycorrhizae and grant them entry. In essence, these mycorrhizae and plants hold a chemical conversation. Phylogeneticists – scientists who study the evolutionary history of organisms – have found that the earliest land plants inherited those genes from their charophyte ancestors, which had (and still have) the ability to exchange signals with fungi.

Any microalgae trying to survive on land also had to withstand constant exposure to drying air. Charophytes' mucilage coating, which evolved for protection from UV light, would have temporarily kept them from desiccating, but mucilage dries out over time. Fortunately, microalgae, including charophytes, also make lipids, which are long chains of fatty acids and include oils, fats, and solid waxes. Modern semiaquatic and terrestrial charophytes have a lipid-based layer that inhibits water loss, which undoubtedly proved useful in their transition to land. Plants' cuticles also employ waxy cutin and suberin as waterproof coverings, and the genes that build these substances are repurposed genes inherited from microalgal forebears.*

By about 500 million years ago, certain charophytes had fully adapted to life on damp land and became Earth's first plants, the liverworts. (Liverwort – such an unattractive name for such an

---

*No red algae ever made it to land. Why not? Genetically speaking, they weren't prepared to make the trip. Scientists posit that during a period of environmental stress long ago, red algae shed many genes in order to save energy and survive. (Even today, a common red seaweed has only two-thirds the number of genes as a common green microalga.) Green algae had more – and more diverse – genes, which meant they could repurpose some to meet the new demands of life on land while maintaining core functions. Red algae had no genes to spare.

important plant. *Liver* comes from their roughly triangular shape and *wort* is from the Old English word *wyrt,* for root.) Today, the smallest and most primitive of the liverworts grow horizontally against the earth and are only centimeters in circumference. While most liverworts have a thin cuticle to protect them from drying out, and their rhizoids are colonized by mycorrhizae, they never evolved the leaves, stems, and vascular systems that more advanced plants use to distribute water and sugars throughout their bodies. In that way, liverworts are more like their microalgal antecedents.*

Gardeners tend to be irritated when they find a patch of liver-

---

*What makes liverworts plants and not algae is how they reproduce. All cyanobacteria and many algae reproduce asexually, by division or budding, but some algae reproduce sexually. In sexual species, parent algae release female and male gametes (germ cells that have only a single set of chromosomes) in the water where, with luck, they encounter each other. The resulting embryo (a *zygote*) is on its own from conception. But plants, including liverworts, harbor their female gametes inside a uterus-like structure (an *archegonium*). The male gametes make their way there and the zygote is nurtured, at least for a short time, within the archegonium. Because liverworts and all other land plants nurture their embryos, they are classified as a subkingdom called Embryophyta.

Liverwort

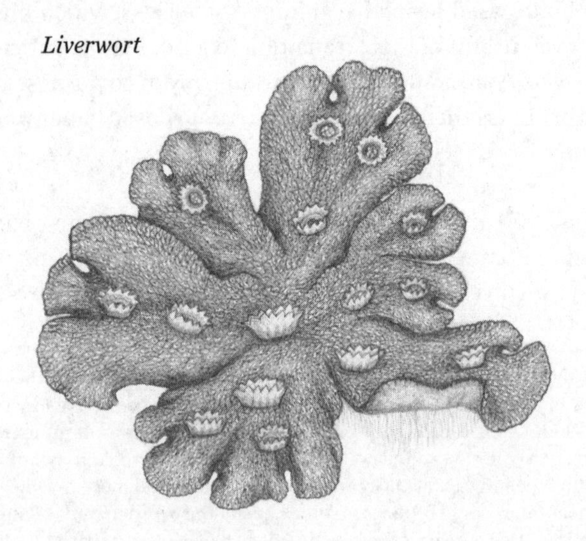

worts in their damp and tree-shaded lawns. Liverworts take hold quickly, and it's true they can block the sunlight that grass seed needs to germinate. They're tenacious creatures, too: if you pull them up and dump them in a compost pile, they'll take hold, grow, and send out spores to repopulate your lawn. But I say celebrate the liverworts; they were intrepid pioneers. And there would be neither lawns nor trees without them.

5

Land Ho, Going Twice

There was another path for algae to land, a path that led to an ingenious — and highly successful — new life form.

In 2005, scientists uncovered microscopic fossils in the Doushantuo Formation in South China that date to between 635 and 551 million years ago. Among the fossils in the shallow, subtidal area were spherical green microalgae, or perhaps cyanobacteria, wrapped in fine nets of filaments. The filaments, called hyphae, belonged to microscopic fungi.

Fungi, a kingdom that today includes molds, mushrooms, and mildew, are the world's demolition experts, decomposing both organic matter and rock with enzymes they excrete from their hyphae. In the Doushantuo Formation, however, the fungi embracing the algae were not deconstructing them; the algae show no signs of damage. Instead, they were stealing some of algae's sugars. Possibly, the fungi were providing a service to the algae in exchange, by protecting them from dehydration when the tide went out. It may be that such collaborations also existed on dry land, and that algae and fungi had struck up a permanent relationship even before liverworts evolved.

In the modern era, the organisms arising from partnerships between algae and fungi are widespread. We know them as lichens. Today, lichens live on tree bark, rocks, bare ground, and nearly any other solid surface. Some, for example, spend their lives adhering to barnacles; one tiny species clings exclusively to driftwood. They're often found on gravestones. There are at least fourteen thousand species of lichen on Earth, and they cover at least 6 percent of our planet's land surface. That's a lot of lichens, but we rarely notice them and often mistake them for something else.

Some look like moss or mold; some seem to be a carpet of crepey, surface-hugging leaves; others appear to be mere stains—black, yellow, orange, gray, blue, or green—on smooth bark or a boulder or even plastic.

My office window looks out on an old Atlas cedar. For many months after we moved in, I worried about the tree. It must have been beautiful once, but its needles were sparse and small, and crinkled gray-green lichens were particularly prolific on the lower, most threadbare branches. I suspected the lichens were harming the cedar and finally called an arborist to see if I should treat what looked like an infestation. But when he came and examined the patient, he exonerated the lichens. While fungi can penetrate the bark of a tree and siphon off its sugars, lichens attach only superficially. We notice them more often, he said, on the branches of old or dying or dead trees for two reasons. One is simply that they are easier to spot through leafless branches; the other is that rotting branches hold more moisture than healthy ones, making them more hospitable sites. (My tree's real problem, he pointed out, are sap-sucking woodpeckers, as evidenced by the many necklaces of holes they've drilled around the trunk.) Although lichens can be a sign of a tree's age, they're harmless.

Whether a lichen looks like a smear of paint, a bunch of crumpled leaves, or a collection of minuscule shrubbery, they all share a basic structure. The microalgae are sandwiched between a top and bottom layer of interwoven fungal hyphae, something like Christmas ornaments packed and cushioned all around in shredded newspaper. The fungi protect their algae from physical damage and excessive ultraviolet light, and they may pass along mineral nutrients. In exchange, the fungi's hyphae penetrate the algae to access their sugars. In some species, the fungi also manufacture toxins that protect both partners from grazing animals, such as reindeer and caribou. About 10 percent of lichens also involve cyanobacteria, either in place of or in addition to microalgae. A threesome is particularly advantageous: the cyanobacteria contribute all-important fixed nitrogen to the ménage.

• • •

A lichen is a curious creature. It is a unique chimera with a thallus that neither acts like nor looks like any of its component organisms. Even though we humans contain ten times more bacterial cells than we do human cells, we're still *Homo sapiens*. Not so with a fungus that takes up with an alga: both parties' identities vanish in the merger, and once formed, they can never be separated. Is the lichen relationship equally beneficial to both partners? The fungus usually cannot live without its photosynthesizing mates, but most of the algal and cyanobacterial species can and do live contentedly without fungi. Even so, I doubt the algae yearn to breathe free; algae in lichen can live in terrain that would otherwise be off-limits — on a sunny rock face, for example, or a wind-scoured roof. It's like marriage; only the spouses know for certain who contributes what to make the relationship work.

We do know that plants would ultimately benefit from the alliance. Without lichens, the first plants with true roots — ferns that evolved about 410 million years ago — would not have found a place to call home. Most plants grow in soil, and lichens, which scientists estimate had been land inhabitants for at least 100 million years before ferns arrived on the scene, were particularly well suited to the business of soil creation.

Crustose lichens — the simplest and oldest lichen type, the ones that look like patches of paint or chalk — probably spearheaded the operation. Now, just as 500 million years ago, a lichen's first step is to grip a rock surface with its myriad minuscule hyphae, which find purchase by penetrating the microscopic spaces and fractures between mineral crystals in the rock.

Weather controls the second step. Because lichens don't have a cuticle to protect them from desiccation, they dehydrate in dry weather, contracting and going into suspended animation. When rain or fog or high humidity moistens them again, they expand and resume photosynthesizing. As they go through these cycles of contraction and expansion, they gradually cleave apart the rock beneath them. At the same time, they also leak acids that scour minerals and trace metals from the rock. As a result, rocks covered by lichens disintegrate ten times faster than they would otherwise.

Finally, lichens' hyphae unceasingly probe the loosened microscopic rock particles to find more solid ground. As they do, the particles become caught up in the almost gelatinous hyphae. When lichens — or pieces of them — die, they leave behind a mix of mineral grains and the lichen's organic compounds. Voilà: soil. Without lichens, Earth would be missing earth. You've got to love lichens.

Or, at least you've got to appreciate how useful they are. Many lichens contain usnic acid, which has antibiotic properties, and lichen extracts have long been used in traditional societies to treat wounds and infectious diseases. Lichens also make excellent brown, purple, and red dyes; until recently, Scottish Harris tweed was colored with lichens. Litmus paper, used to determine the pH of liquids, contains lichen extracts that change color depending on the liquid's acidity or alkalinity. And perfume companies use tons of lichen every year to create "mossy" notes in their products.

Most lichens are too acidic for humans to eat without serious indigestion, but a few species, properly prepared, are edible. For at least a thousand years, Icelanders relied on a lichen misleadingly called Iceland moss for sustenance during the winter months. Farmers and villagers would collect it late on summer nights, after the dew had softened it, making it easier to pick from rocks. After spreading it in the sun to dry and cleaning it, they stored it in barrels and woven bags for leaner times. These lichens are 70 percent polysaccharides — those long chains of sugar molecules — and during the long winters when food was scarce, people boiled them in water for tea, with milk for soup, or with milk and wheat to make a porridge. As noted in *The Dispensatory of the United States of America,* in 1918, "The gum and starch in the moss render it sufficiently nutritive to serve as a food for the Lapps and Icelanders only after having partially freed it from the bitter principle by repeated maceration [i.e., steeping]." Note the "partially"; lichen porridge was a food of last resort.

Lichen species rich in polysaccharides are also candidates for fermenting into alcohol. In the 1800s, Scandinavians began brewing a lichen liquor, and by 1869 there were seventeen lichen distilleries in the region. Today, it's a lot easier to grow potatoes to make vodka than to find and pick lichens for liquor (to say nothing of the environmental damage involved), so lichen distilleries are a thing of the past. I discovered, however, that Íslensk fjallagrös, an Icelandic company, makes schnapps with lichen. It isn't available in the US, but if you're ever in Reykjavík, have a glass and give a toast to algae.

While lichens are not often on the human menu, reindeer and caribou couldn't survive without them. A species called reindeer moss makes up half their winter sustenance. Reindeer moss is a soft green-gray shrubby species that covers miles and miles of tundra in northern Europe, Asia, and North America. Reindeer and caribou are ruminants whose four-chambered stomachs produce an enzyme called lichenase that decomposes a lichen's acids. Certain indigenous tribes in the Arctic, after killing a reindeer or a caribou, may remove its stomach and eat it with the semi-digested lichens found inside. The Boreal Forest website of Lakehead University in Ontario, Canada, reports that such a lichen "tastes like fresh lettuce salad," but I have my doubts. I also read that some tribes mix a reindeer's stomach with the animal's fat and blood plus meat scraps and liver to make a pudding that is eaten fresh and warm, but uncooked. (No word on the taste.) If you're in Copenhagen and are an adventuresome diner, check out Noma, a restaurant that sometimes offers the lichen on its menu.

I'm not inclined to eat lichens, but there's something about them — these quiet shape-shifters, wallflowers at the biosphere ball — that appeals to me. They're omnipresent but unknown; fragile yet durable, both unassuming and heroic. Newly alert to their heritage, I look for them.

6

Looking for Lichens

It turns out that in my suburban neighborhood, the lichen pickings (not that I would pick them) are slim. I find only pale blue powdery types growing on the smooth bark of birches and crumpled-leaf, silvery green sorts like the one on my cedar. I'm impressed by their ubiquity, but I don't find the variety I'd been hoping for.

Which is why, on a trip to San Francisco, I am pleased to wangle an invitation to join a guided lichen walk in Point Reyes National Seashore Park in Marin County, forty miles north of the city. I drive from my downtown hotel and, as I near the park, I follow a road that skirts Tomales Bay and turns off onto a one-lane road that winds up through the hills. Together, the lane and I bend and rise a thousand feet, first through grasslands and then hills covered in bay bushes, scraggly coyote brush, pines, and broad live oaks. Some of the trees are festooned with light green bird's nests of fibers — I'm surprised to see Spanish moss so far from the South. Finally, the lane peters out in a gravel parking lot where I meet the group. Most of the others have already gathered when I arrive, and I find my fellow hikers congratulating themselves on our luck with the weather. The morning fog has just burned off and the day is now warm under a spotless blue sky.

A perfect day for lichening, I'm told. Lichens are brightest when they're damp; after a day or two without rain or fog, they dry out and their colors fade. Today, lichens will be saturated with the morning's mist, primed for as full a display as these unpretentious creatures ever put on.

Our guide for the day's walk, Shelly Benson, is a slender young woman with warm brown eyes and, beneath a baseball cap, brown hair pulled into a ponytail. She is a lichenologist and president of the California Lichen Society and, after I introduce myself, she offers me a plastic magnifying glass on a chain — birders accessorize with binoculars; lichen buffs bear loupes. Then she rounds up her charges and reminds us about sunscreen. We start down the path to find lichens.

And immediately stop.

The first tree we come across, a live oak, is home to half a dozen markedly different species. We crowd around the trunk, taking turns peering through our loupes like jewelers inspecting a tray of gems.

But Shelly diverts our attention, pointing above our heads to a branch draped in that hanging moss I'd seen on my drive into the park.

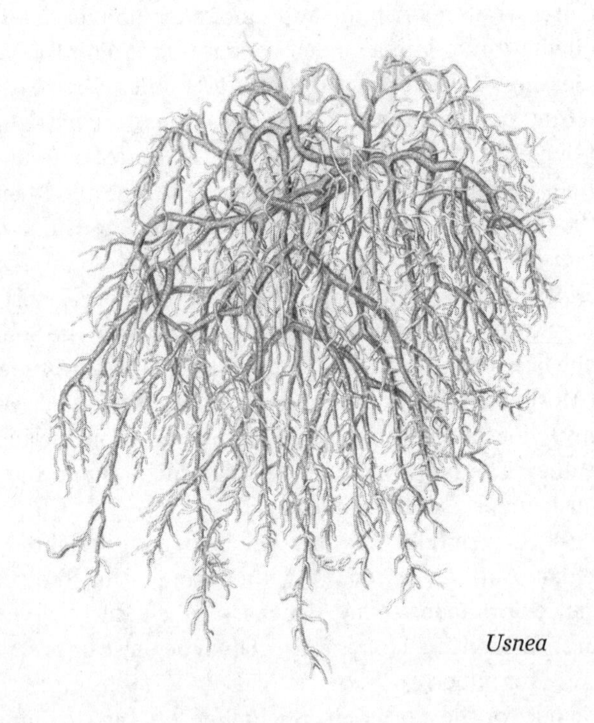

Usnea

"Does anyone know what this is?" she asks. Two decades of schooling tell me this is too easy. I keep quiet.

"Old man's beard," a female voice behind me volunteers.

"Usnea," says the middle-aged man next to me wearing a shirt and shorts that are more assemblages of pockets than clothes.

"You're both right," Shelly says. "It *is* an *Usnea*. And I'm really impressed that no one said Spanish moss."

There are a few chuckles, whether from embarrassment (that would be me) or amusement that anyone could possibly be so daft.

"I call it rubber band lichen," she adds, as she reaches up to pinch off a strand of it and pulls both ends. The strand stretches; when she releases it, it recoils. With a thumbnail, she scrapes off its outer covering so we can see the thin white cord inside. The cords of the *Usnea* owe their elasticity to fungal tissue. Should you ever wonder whether you're looking at Spanish moss or *Usnea*, the elasticity of the lichen is diagnostic.

Usnea is a fruticose lichen, a word that, despite its pronunciation, has nothing to do with fruit. *Fruticose* comes from the Latin for *shrub,* and these lichens often resemble miniature tumbleweeds and attach to a substrate at a single point as most plants do. In fact, model train sets once used fruticose lichens to stand in for shrubbery.

Say *lichen* and most people think of the foliose type, and on the bark in front of us, Shelly points out a good example: a yellow-green flat rosette composed of what look like fingernail-size leaves, buckled and wrinkled. This is a Greenshield, and it is quite common across the globe, the sparrow of the lichen world, if you will. Like all foliose lichens, it grows outward from its edges. Sometimes the center crumbles and blows away, leaving a vacancy that the lichen refills.

Which brings up the question of reproduction. How do these chimerical entities make more of their bipartite or even tripartite selves? Imagine the difficulties involved when two or three entities, each with different reproductive strategies, are bound indissolubly together. No surprise: lichen sex can be complex.

For this reason, many lichens stake out new territory asexually. When a fragment detaches and blows away, if it lands in a similar location, it attaches and starts growing as a new individual. *Usnea* species are well suited to this asexual approach. A piece of a filament breaks off and, like a cutting from a plant, establishes itself on another branch.

Many lichens go about asexual reproduction more deliberately, though, and the Greenshield is one. Shelly urges us to look very closely at it, which means, one after another, we hold a loupe up to an eye while air-kissing the tree. "If you look really carefully, can you see bumpy, rough areas on the thallus? Those are soredia. If you touch them, they'll feel dusty. Those are actually tiny balls of algae wrapped in fungal hyphae." Soredia emerge from what look like pustules on the surface or edges of a thallus. After the pustules open, wind or animals distribute the lichens' vegetative spores.

But lichens do get it on, and conveniently enough for Shelly and for us, a sexually reproducing species is growing near the base of a neighboring tree. The lichen in question is the lipstick or pin lichen, *Cladonia macilenta,* a startling but common sight (if you're on the lookout for lichens) along the West Coast and in the eastern half of the US. The thallus looks like rough, low-growing moss or the plastic scrubber side of a kitchen sponge. Rising from its surface are dozens of inch-tall, pale green "pins." The heads of the pins are daubed bright red, as if they've recently pricked elfin fingers.

The tiny red balls are ripe fruiting structures called apothecia. Like the more familiar fruits we eat, they hold reproductive cells — in this case, fungal spores. Some fungi form cloned spores that, without further ado, grow into new fungi if they land in a propitious spot. But most of the fungi that lichenize (that is, join with an algae species to form a lichen) are Ascomycetes, and members of this phylum often reproduce sexually. Sex in a fungus involves two rounds of chromosome division and a fusion between two structures within the fungus before the spores can go skipping off from the apothecia into the ether.

And that's the easy part, Shelly explains. To become a lichen, a fungal spore has to land on an alga. And not just any old alga: It has to touch down on a member of the one and only species it can lichenize. What are the chances a fungal spore will find its beloved algal mate? Well, it certainly helps that fungi produce millions or even billions of spores. (For example, a giant puffball mushroom, which can grow as big as a footstool, shoots several trillion spores into the air.) It helps, too, that the spores are so lightweight that they rise in the air and easily travel dozens of miles. Nonetheless, it's no wonder that, when it comes to reproduction, lichens often have an asexual plan B.

Our sex-ed lesson complete, we mosey on along the trail, breaking into twos and threes as we go. The terrain is gentle, the air is fine and fragrant with warm soil and a hint of the sea a few miles beyond and below us, but our progress is slow — all along the path are lichens. It's a lichen bazaar, and the more closely you look, the more you see. Here are pixie cup lichens, which sport what look like tiny, gray-green golf tees. There's a leafy, blue-gray shield lichen covered in pustules and soredia. One species of rim lichen looks like a sprinkling of olive-green buttons; another looks like the surface of cauliflower. A tree-hair lichen could be a swatch of coarse hair clipped from a horse of a different (green) color. A fallen branch is so encrusted with a mosaic of lichens that it's a topographical map, each different-colored patch representing a different country, some hilly, others Death Valley flat.

At noon, we stop for lunch in a clearing near a fence. I untie my jacket from around my waist, spread it on the ground, and use a white plastic fence rail as a backrest. I've taken a seat near Shelly, who draws the group's attention to my fencepost. On it are dark hairline scratches. I assume they're dark because they're filled with dirt. Not so; black crustose lichens are sheltering in these minuscule crevices.

Across the small clearing, a fellow hiker goes to rest her hiking pole against a boulder and discovers at its crest a small, vivid-orange tumbleweed of a lichen. It's a solitary brooch on the uniform gray surface of the rock. Shelly is particularly delighted when she investigates. This is golden-hair lichen, she tells us. "This lichen grows only in coastal areas and is relatively uncommon. I think it should have conservation status. Because there is so much development along the coast here, there just isn't much habitat left for it. It needs protection."

We gather around it reverentially and take cell phone photos.

After lunch, we turn around and head back to the parking lot. We may have only walked about a mile, but I've come away with new eyes (and a loupe) for my walks in the woods. Now, though, I'm curious about Shelly, wondering how she fell for lichens and whether she can make a living from her passion. There are all kinds of jobs for botanists and horticulturalists; crops cover half the Earth's land surface, and nursery, greenhouse, and related industries are a $20 billion business in the US alone. We need mycologists: mushrooms are a major crop, and fungi, including molds, can be serious pathogens in agriculture. Algae and seaweeds are a big business, too, so we need phycologists. But no one is cultivating, or battling, lichens. How is it possible to earn a living as a lichenologist?

"I'm not entirely sure there is a way," she laughs later, as we sip our coffees in the nearby town of Inverness, "but I'm trying."

Shelly did not set out to be a lichenologist. She grew up in the small town of Yelm in the shadow of Mount Rainer in Washington, went hiking often as a child, and studied botany and ecology as an undergraduate. In the summer before she graduated, she took a job at the research station in the Wind River Experimental Forest in the Cascade Mountains of southern Washington. Researchers have been monitoring plots in that forest since 1909, studying the 400-year-old Douglas firs and other trees to learn how to prevent or control fires and how to manage a diverse ecosystem, especially in an era of climate change. In 1994, a construction crane was installed so scientists could study the vegetation in a columnar plot that stretches from the forest floor to the top of the canopy. In the summer of 1998, Shelly joined a grant-funded project and spent time in the crane's gondola. Her particular task was identifying lichens and analyzing how the vertical gradient affects the populations. At summer's end, she was invited to become a graduate student in lichen ecology at the University of Northern British Columbia in Canada.

"With an offer of a full ride, I thought, sure, why not?" she recounts cheerfully, and then adds more soberly, remembering, "The university is nine hours north of Vancouver, so that was a big dose of winter."

Her graduate studies completed in 2001, she began piecing together a living as a lichenologist, jumping from job to job, following short-term and seasonal research projects at universities and the US Forest Service. All the projects she worked on were funded by grants, and the grants rarely lasted longer than a year or two. For a while, in search of a steadier paycheck, she signed on with a consulting company in California to do environmental assessments required under the state's Environmental Quality Act.

"Every time a company needs to disturb soil, for building a culvert or clearing a utility right-of-way or developing a shopping center, there has to be an environmental impact assessment to see if there are any rare plants. I was used to working in beautiful natural areas; then, I found myself doing surveys of highway ramps and vacant lots. I didn't enjoy that, so I quit to see if I could get some lichen work going."

The most meaningful work, Shelly believes, is expanding the use of lichens as monitors of air quality and climate change. Lichens rely on nutrients they gather from the air, and they indiscriminately absorb and concentrate gases and particles that float past or fall on them in rain or fog. In the mid-nineteenth century, naturalists began to observe that lichens disappeared around cities wherever industrial smoke grew dense. It was Finnish botanist William Nylander who, in studying the disappearance of lichens around Paris in the late 1800s, realized that their relative numbers are a gauge of type and concentration of pollutants in the air.

Take as examples *Usnea* and *Teloschistes*, species that are sensitive to sulfur and nitrous oxides. Sulfur dioxide, emitted primarily from burning coal and oil and from industrial processes like nickel smelting, becomes acid rain, which kills plants and fish in lakes and ponds and damages human lungs. Coal-burning power plants and vehicles emit nitrogen oxides that react in the air to form corrosive nitric acid. These two lichens are the proverbial canaries in the coal mine, only instead of warning of explosive levels of methane or carbon monoxide, they warn of perilous amounts of sulfur or nitrogen compounds. *Usnea* and *Teloschistes,* a bright orange sunburst of a lichen, die when levels are high, while crustose lichens like dusty gray-green *Lepraria incana* become more prevalent. Surveying these and other indicator species gives scientists a way to follow the impact of pollutants on the environment in a nuanced way over time, and it's much less expensive than deploying an equivalent number of mechanical air samplers.

Sometimes the lichens' news is good: their return to Sudbury, Ontario, after the local nickel mine installed sulfur-scrubbing equipment proved the scrubbers were effectively remediating the environment. Sometimes the news is mixed: surveys of *Usnea* in England and Ireland reveal where and when the Clean Air Acts have and haven't improved air quality. Urban lichen monitoring in Great Britain demonstrates how public green spaces clean up air, as well as how traffic bottlenecks sully it. Sometimes the news is all bad. After the 1986 Chernobyl reactor disaster, scientists sampled lichens, which readily absorb radioactive contaminants, to determine the volume and paths of nuclear fallout. Because lichens are also very long-lived, those exposed to Chernobyl fallout continue to be a source of radiation. To this day, reindeer herders in central Norway have to release some of their captured animals because the lichens the reindeer have eaten would make their meat too dangerous to consume.

In the US, the Forest Service has been collecting lichen data in national forests for nearly fifty years to assess how air pollution and climate change are affecting ecosystems. Shelly hopes to deepen our understanding by expanding monitoring to sites in nonfederal public areas. Lichens are some of the land's oldest and most venerable occupants, critical to the evolution of terrestrial life. It is fitting that scientists enlist them to track the continuing health of our planet. It pays to listen to lichens.

SECTION II

GLORIOUS FOOD

1

Brain Food

I'm watching a furry black chimpanzee, nearly as big as an adult human, plant herself on a cluster of gray boulders by a rainforest stream near Bakoun, Guinea, in western Africa. She grasps a long, straight branch and, holding it in both her hands, deftly maneuvers its tip to prod the bottom of the stream. Fifteen seconds pass and she lifts the pole above the surface to inspect its far end. Whatever she is expecting, she hasn't gotten it, and she lowers it and continues to probe. A little more time passes, and again she lifts it out of the water. Draped over the far end are skeins of dripping seaweed. She slides her "fishing rod" to shore — hand to hand just as you or I would — grabs the seaweed, and devours her lunch.

I am able to spy on these chimp gourmets thanks to a video from the Max Planck Institute. The video delights me, and I'm inspired to watch more clips of chimps of all sexes and ages fishing for seaweed. I then read that this untutored behavior has been spotted in other troops in Guinea and in Congo, as well. This is especially intriguing given what I've been learning about the role of seaweed in the evolution of our species.

One of the hot topics debated by paleoanthropologists today is what led one branch of early primates to diverge from the others and eventually become us, *Homo sapiens*, with brains three times as large as these Guinean chimps' brains. It's not as if having a large brain has made us, among all other primates, uniquely capable of surviving and prospering. Apes, chimps, orangutans, and baboons have been successful for millions of years — far longer than we have. So, what first sent our ancestors on a different cognitive path? How did the earliest hominins develop a slightly larger brain than their predecessors'? And how did their descendants continue to evolve ever-larger brains?

To understand, we first have to meet our ancestors the Australopithecines. They were four- to five-foot-tall, hair-covered, bipedal hominins with simian faces who lived in eastern and southern Africa from four million to three million years ago. Their more apelike predecessors had evolved in dense forests, but at this time the climate was growing drier, shrinking the forests, and creating in their place savannas with isolated stands of trees. The more open plains were punctuated by strings of lakes and their surrounding marshes, which had been created as a vast subterranean cleavage gradually opened to become the Great Rift Valley. The small-brained Australopithecines had physically adapted to the transformed environment. They retained the long arms and curved fingers of their entirely tree-dwelling forebears, but their lower bodies had changed: the orientation of their pelvises had shifted so they were able to walk upright, albeit not with the same stride and ease that we do. They modified their behavior, as well. They may have slept in trees at night, but they took advantage of their mobility to forage at nearby lakeshores during the day. The food they found there supplemented their ancestral diet of leaves, fruits and nuts, ants, grubs, and the occasional small animal.

Like modern apes, Australopithecines would have been comfortable wading in shallow water. We know by the chemical signatures in their fossil bones and wear patterns on their teeth that they ate tough sedges and reeds that flourished along the shoreline. Not likely fastidious eaters, they would have ingested the small snails that clung to those plants, and scooped up small, transparent crustaceans and fry, just as bonobos do today. These foragers would have inevitably stumbled across the nests of shorebirds and crocodiles and eaten their eggs. They may have plucked freshwater clams and mussels out of the shallows, grabbed frogs and turtles, and gathered seaweed as the Guinean chimps do today. Two types of fish, catfish and cichlids, would have been easy to nab, thanks to their breeding and hibernation patterns. Catfish seek out shallow waters to spawn; in the dry season, they burrow into mud to survive until the rains return. Cichlids brood their fry in the shallows.

By foraging in the marshes, certain Australopithecine species changed the evolutionary trajectory of hominins. Brains are energetically expensive; while they are only 2 percent of our body's mass, our brains use as much as 20 percent of our energy. A forest-based diet simply didn't have sufficient calories to support bigger brains. Animal meat could have provided extra calories, but while Australopithecines might have enjoyed an occasional carrion treat, large animal carcasses would have been rare finds. Hyenas and vultures — far fleeter and with a sharper sense of smell — would have likely gotten to the table first. Nor were Australopithecines able to hunt animals: their feet were not suited to running swift beasts to exhaustion, and they lacked both weapons for killing large prey and stone tools for butchering them. Along the shore, however, the extra calories were easy pickings (crocodilians and other lurking carnivores notwithstanding). Even pregnant and nursing females could gather these foods without tools or traveling.

As important, along the shoreline Australopithecines were also inadvertently consuming certain minerals and fatty acids — collectively known as brain-selective nutrients — that are indispensable in building brain cells and neuronal networks. One of the most important brain nutrients is iodine, a mineral critical for human brain development because it is the chief ingredient in thyroid hormones. Thyroid hormones are made in the small, butterfly-shaped thyroid gland at the front of the neck. Without enough iodine, humans cannot make a sufficient amount of these hormones, and without enough thyroid hormones modern humans become lethargic, unable to concentrate, and our short-term memory fails. Insufficient iodine results in an enlarged thyroid gland, a condition called goiter that, in the worst cases, looks like a pair of apples under the skin. Severe deficiency can lead to coma and death. Thyroid hormone deficiency in pregnant and lactating mothers retards brain development in fetuses and neonates, and a severe lack *in utero* leads to the physical stunting and brain malformation once called cretinism. Even mild iodine deficiency during pregnancy can permanently prevent a child from developing his or her full intellectual capacity.

Iodine is, unfortunately for us, scarce on land. Volcanoes blast it into the air, but since 70 percent of Earth is covered by oceans, most of it falls into water. Even the iodine that does settle on land eventually ends up in the oceans, thanks to erosion. Before the iodization of salt in the early twentieth century, people living in mountainous Appalachia and the glacier-scoured areas of the American Upper Midwest and Northwest were particularly susceptible to goiter and even cretinism. These regions were part of the "goiter belt"; the prevalence of the disease was recorded at 64.4 percent in parts of Michigan. In some regions of Switzerland, before salt was iodized in 1922, almost 100 percent of schoolchildren had large goiters and up to 30 percent of young men were deemed unfit for military service due to the disease.

It doesn't take living on a glacial plain or in the mountains to put one at risk of iodine deficiency. Simply living far from aquatic foods can be a problem. While plants can take up iodine dissolved in groundwater, the amount they accumulate in a season is minimal. When crops are grown and harvested again and again on the same land, as has been the case in many parts of the world, iodine levels fall farther. In the 1970s, 370 million Chinese people lived in iodine-deficient areas and, according to a Chinese journal, "Deficiencies manifested various ways, including goiter, cretinism, endemic mental retardation, and decreased fertility rates." (After the Chinese government mandated iodized salt, the prevalence of goiter fell to half the 1970s levels by 1996.) Frequent floods leach the iodine out of Cambodia's soil; in 1997, nearly a fifth of Cambodians suffered from goiter. The incidence of goiter has also long been widespread in inland areas of India, Central Asia, and Central Africa. For those of us in the developed world who have grown up with iodine-supplemented food, it is difficult to appreciate how common iodine deficiency once was and how often it harmed full cognitive development. Even today, almost 30 percent of the world's population lives in iodine-deficient areas without access to seafoods, and the World Health Organization estimates that iodine insufficiency remains a health concern for hundreds of millions of people around the globe.

The recommended daily allowance of iodine for adults is 150 micrograms, and today a half teaspoon of iodized salt provides as much of the mineral as most of us need. (Lactating mothers need twice as much.) Because most of the world's iodine is in the oceans, the best natural sources of it are marine algae and seaweeds. A serving of kelp is astonishingly rich in the mineral, containing 2,000 micrograms or more, and other seaweeds, if not quite so amply endowed, are also excellent sources. And because fish, mollusks, and other marine animals either eat algae directly, eat zooplankton that eat algae, or eat smaller fish that eat zooplankton, they are all excellent repositories of iodine, as well. A serving of haddock has 250 mcg and salmon has 150 mcg; a portion of scallops has 125 mcg and shrimp has 25 mcg. Although freshwater fish and shellfish generally have only about 10 to 20 percent of the iodine content of marine species (river and lake water has less iodine than saltwater), they still have significantly more than most terrestrial plants, which often have only trace amounts. By dining along lakeshores, Australopithecines and other early hominins had access to significantly more of the mineral.

Algae also provided another ingredient essential for evolving larger brains: the polyunsaturated fatty acid called docosahexaenoic acid or DHA. One of two kinds of omega-3 oils, DHA is found in the membranes of brain cells. It's concentrated in the synapses of neurons and makes up a substantial portion of the brain's physical matter — the bricks, so to speak, of our brains. It also triggers more than a hundred genes that are important in brain growth and development in fetuses and infants, and is needed for the production of transthyretin, which carries thyroid to the brain. Ancient hominins snacking along the shorelines inevitably consumed more DHA than their strictly forest-dwelling predecessors.*

Some scientists (most prominently Stephen Cunnane, professor in the Department of Medicine and University Research Chair in Brain Metabolism at the Université de Sherbrooke in Quebec, and Michael Crawford, director of the Institute for Brain Chemistry and Human Nutrition at the Imperial College London) posit that several million years ago, with just a little more iodine and

---

*While our bodies can convert alpha-linolenic acid (ALA), an omega-3 oil found in plants, into DHA, the process is limited.

DHA in their diets from visits to the marshes, early hominins evolved brains that were just a little bit larger than their predecessors'. While a species' absolute brain size is not a stand-in for intelligence—just think of clever crows and parrots—a measurement called the Encephalization Quotient or EQ, which is the ratio of brain size to body size, is more meaningful (although even it still doesn't tell the full story on comparative intelligence). In a system of EQ that pegs modern humans at 100, modern gorillas and orangutans weigh in at 25 and 32, respectively, and *Australopithecus africanus* measures 44. The latter's direct descendants in the Rift Valley, *Homo habilis* (the earliest members of the *Homo* genus who lived roughly 2.4 million years ago), had an EQ of 57. *Homo erectus* lived from about 1.9 million years to 150,000 years ago, and individuals who lived at the end of that period had EQs of 63.

What processes could translate more iodine and DHA into brain capacity? Start with the fact that the nutrition a third-trimester fetus and then a nursing child receives through its mother greatly influences that individual's brain capacity. If a pregnant or nursing hominin regularly ingested slightly more iodine, her thyroid gland would be stimulated to produce slightly more thyroid hormones, which would pass through the placenta to her fetus. If the fetus happened to have, simply by natural variation, a few more thyroid hormone receptor molecules than average, then the processes of neurogenesis—brain cell proliferation and differentiation—would have been slightly enhanced. If infants with a slightly greater ability to make use of iodine tend to survive and live to procreate at a greater rate, then the continued presence of iodine will, over generations, produce individuals with larger EQs.

The abundance of DHA in the shore-based diet was similarly beneficial and likewise became a requirement for normal infant development. In fact, hominin reliance on a supply of DHA led to a fundamental change in the pattern of growth in human fetuses and babies. Unlike all other animal neonates, newborn humans already have—and continue to accumulate—a layer of "baby fat." This fat, stored under the skin, is a source of stored calories. It is also three to four times richer in DHA than adult fat and ensures that a human child has enough DHA for its rapidly developing brain. The mother passes along DHA to her fetus *in utero* and to her infant through breast milk.

Clearly, we *Homo sapiens* ultimately benefited from the fact that our long-extinct ancestors consumed more food from along shorelines. But, when — and where — did *Homo sapiens* with fully modern, EQ 100 brains evolve? It certainly must have happened at a time and place where algae-concentrated DHA and iodine were on the menu. In other words, along the shore and, most likely, along the ocean shore.

While there continues to be much debate about the origin of our own species, the current consensus is that we evolved sometime around 230,000 years ago in eastern and southern Africa. At the time, the climate was mild, rainforest extended from the east to west African coasts, and crocodiles and hippos trundled through what is now the Sahara Desert. But another glacial age soon set in, cooling and drying the continent dramatically. Nearly all of Africa north of the equator became a desert, making much of the continent uninhabitable. Most tribes of *Homo sapiens* perished, and our species neared extinction; only a cluster of humans, including a few hundred women of childbearing age, survived. Most paleoanthropologists now agree that 165,000 years ago, these Africans found a refuge and lived through catastrophic climate change to become the Adams and Eves of all modern people.*

Where was that refuge? Dr. Curtis Marean, associate director of the Institute of Human Origins at Arizona State University, makes

---

*When our *Homo sapiens* ancestors spread beyond Africa between 70,000 and 54,000 years ago by following the coastlines and rivers, they eventually encountered Neanderthals and Denisovans, species descended from *Homo erectus* that had left Africa 1.8 million years ago. Through interbreeding, roughly 4 percent of the *H. sapiens* genome has non-*sapiens* genes, but because of the genetic bottleneck in Africa, modern humans are remarkably similar genetically.

So, why do we have such variable facial features? We are a highly social species and, unlike other animals who distinguish one another chiefly by smell or vocalizations, we use our highly developed visual sense. Humans are extraordinarily adept at recognizing faces (we have areas of the brain devoted to that function), and our species — as well as Neanderthals and Denisovans — evolved so that individuals are unique and easily recognizable to one another.

a convincing case that the coast near Cape Agulhas at the southern tip of South Africa was our sanctuary. Marean has spent twenty years directing the excavation of several cliff caves at Pinnacle Point just east of the Cape. About forty archaeologists, funded by the Smithsonian Institution, the National Science Foundation, and others, are unearthing details of the lives of those early *Homo sapiens* and the conditions that saved them.

It turns out that the environment of this part of South Africa was uniquely hospitable for our ancestors. The Agulhas Current, which flows southwest from the Indian Ocean, moderated the Cape's climate, much as the Gulf Stream warms the climate of northwestern Europe (which otherwise would have weather more like Newfoundland's). Moreover, at the Cape the Agulhas Current meets cold, nutrient-rich waters from the Atlantic, and the juncture makes for one of the most productive marine environments on the planet. Dense beds of mollusks – among them brown mussels, limpets, and sea snails – glue themselves to the rocks in the intertidal zone. These bite-size marine animals survive by filter-feeding algae or, in the case of brown mussels, algae plus zooplankton. The Cape coast is also one of the most productive habitats for seaweeds in the world; hundreds of species, including relatives of two eminently edible species, *Porphyra* (the red seaweed we know as nori) and *Ulva lactuca* (sea lettuce), grow on the rocks and hard sandstone, and sometimes – why not? – on the mollusks. For sure-footed early humans, the Cape shoreline was a smorgasbord, uniquely packed with DHA and iodine in amounts far greater than at the freshwater sites their forebears knew in the Rift Valley. Brown seaweeds are so rich in iodine, it would have taken only a scrap to get an adequate dose.

Humans partook of all that was available. Dr. Marean and his colleagues have discovered discarded shells that demonstrate that the residents collected and ate these mollusks. (Today, people drive to Pinnacle Point to collect mussels and seaweed and, when tide conditions are right, they can collect a meal's worth in no time.) It is also reasonable to assume, based on both chimps' and modern humans' fondness for seaweeds, that the area's inhabitants also took advantage of them, either eating them raw or cooking them briefly on hot rocks.

The Cape had another distinct advantage for these early human settlers. The caves are close to the Fynbos, a unique narrow band of arid scrubland that is home to the most diverse array of plants in the world. (There are more species in this one small region than in all of Great Britain.) Many of the plants store their energy as carbohydrates in edible underground tubers and corms. Until the last century, native gatherers used digging sticks to unearth them and, no doubt, their Pleistocene ancestors did as well. All in all, the South African coast wasn't just a place to survive but to thrive, perhaps as our species had never done before.

The well-nourished people at Pinnacle Point had developed advanced brains, as evidenced by artifacts they used. Archaeologists found the earliest worked pieces of ochre, some of which had been intentionally heated to intensify their red color. The ochre was likely used as a body paint, and paleoanthropologists consider it evidence of early humans' ability to think symbolically, conveying ideas like "this is my tribe" or "this is my position in my tribe." They also found intensively flaked pieces of silcrete, a fine-grained rock that can only be shaped after it has been slowly heated to more than 650 degrees Fahrenheit (343 degrees Celsius), a feat that requires at least a rudimentary rock kiln.

Marean points to another indicator of a thoroughly modern intelligence. Today, the Pinnacle Point caves are fifty feet above a rocky coast, but when the first *Homo sapiens* settled in, they were miles from the shore. The settlement was wisely chosen, equidistant from the Fynbos for foraging tubers and from the shore for collecting mollusks and seaweeds. But while tubers were available at any time, shellfish — or at least enough of the small creatures to make the trek to the sea worth the effort — could only be collected at certain low tides. Because of the monthly tidal cycles, those tides occur only about ten days per month and last only for a few hours. It would have taken a high level of cognitive ability to figure out when a group should start the journey to the ocean.

Full cranial volume probably arrived before the fully modern human mind, which involved shifts in the relative size of different brain regions, as well as the density of the connections among them. The ochre, the silcrete, and the reckoning of tides indicate that the humans at Pinnacle Point had taken a leap in cognition and were capable of complex, modern mental processes. Were these achievements related to the surge in the amount of iodine- and DHA-rich seafood they ingested? Whether access to seafood permitted the development of the fully modern, analytic, and creative mind or whether the fully modern mind permitted our ancestors regular and constant access to these nutrients, or whether the two developed simultaneously is difficult, and perhaps impossible, to determine conclusively. What we do know is that the evolution of our species — and our predecessor species — was inextricably linked to consuming brain-selective nutrients consistently and over a long period of time. Without algae in the hominin diet, we would never have parted ways with our less brainy primate relatives.

Algae continued to influence the history of *Homo sapiens*. Scientists now believe that algae were critical to the first settlement of the New World. In 2007, University of Oregon professor Jon Erlandson and fellow archaeologists and marine ecologists collaborated to determine the path that East Asians took when they migrated to the North American west coast. Before Erlandson and his colleagues published their research, most scientists believed that these migrants walked across an Ice Age land bridge between Siberia and Alaska and then trekked down through North America via an ice-free route east of the Rocky Mountains. Now it is understood that massive ice sheets would have blocked an early inland passage. Paleoanthropologists now believe that intrepid immigrants from the Kuril Islands, which stretch between Japan's Hokkaido island and Russian's Kamchatka Peninsula in the north Pacific, came by boat from East Asia, paddling along the shoreline of either a Siberia–Alaska land bridge or the Aleutian Islands to modern Alaska. From there, they continued down the Pacific coastline, reaching southern Chile as early as 18,500 years ago.

That ancient route is now known as the kelp highway. The Ice

Age coastline of the American continents was much lower and more convoluted than it is today, with more shallow bays and inlets. Thick beds of kelp sheltered fish and shellfish, as well as otters, seals, and other marine mammals. In settlements at Monte Verde, located on a creek a few dozen miles from the coast in southern Chile, archaeologists have uncovered hearths that hold the ash remains of nine species of seaweed. The western coast of the Americas was the perfect environment for humans, whose brains depended on iodine and DHA, to flourish.

Around the world, early humans tended to first migrate along the shorelines and only later move inland, far from the algal and marine animal sources of brain-selective nutrients. Some inland dwellers had sufficient access to freshwater fish and eggs to maintain adequate iodine levels, but many did not, and deficiencies would have been widespread. Today, iodized salt generally prevents the iodine deficiency that has handicapped so many over the millennia. As for DHA, long ago our species evolved so that our newborns and breast-fed babies are ensured a supply for healthy brains. Supplemented infant formula now fills any nutritional gap.

Hominin — specifically human — history is intimately tied to algae. But, as we'll see, a continuing supply of omega-3 oils is still important for adult health.

2

Seaweed Salvation

Even without iodized salt, I probably get my complement of iodine and DHA simply because I have long been addicted to a Japanese dish called maki sushi. Maki sushi is a seaweed-wrapped roll containing cold rice and raw vegetable and/or fish that is sliced into bite-size pieces. My favorite variety is made with raw salmon and ripe avocado. I appreciate the rich creaminess of both; the textural similarity between a fish and a fruit pleases me. The dish could be insipid, but the twinge of the vinegar in the rice and the tingle of pungent wasabi save it. Over the decades, I figure I must have consumed, slice by slice, a small school of salmon, bushels of avocados, and a basketful of the roots of *Eutrema wasabi*.

But until recently, in all this blissful eating I never paid attention to the dark-green seaweed — nori — that holds the roll together. I certainly didn't imagine that the nori had any nutritional value, and I didn't focus on its taste. I never thought of the seaweed as anything other than a solution to the problem of keeping rice, fish, and avocado together long enough to travel from plate to mouth. But now, having attentively sampled this particular macroalgae on three continents, I know it is more than an edible conveyance.

In East Asia, nori is eaten at nearly every meal, not only with other foods, but on its own. In Japan and Korea, crisp, crackly, seven-by-eight-inch sheets of nori are often placed on the table in a neat pile, like a stack of cocktail napkins, ready to use to pick up a bite of rice. Sometimes strips of nori, teriyaki- or wasabi-flavored, serve the same purpose. Onigiri, rice balls with a sweet or savory filling (my favorite is salted plum), are nestled in a broad strip of nori and sold everywhere, including the Japanese equivalent of 7-Eleven stores. Cooks across East Asia pulverize the sheets and sprinkle bits on salads, stir them into soups, or mix them with rice. And in Japan, people snack on nori rice crackers the way Americans nibble potato chips.

In the last decade, a wide variety of seaweed snacks — toasted sheets of nori often dusted with wasabi, salt, or other flavors — have staked out territory on the shelves of grocery stores across North America and Europe. You'll find nori snacks in elementary-school kids' lunchboxes. It's a shame that in English we call macroalgae *seaweeds.* A seaweed by any other name — say, *sea vegetable* — would certainly taste sweeter. Or, we could try *sea herb,* a more accurate translation of the Old English *sæwar*.

There are excellent reasons to eat nori in addition to its pleasing taste. Although seaweeds, including nori, are not manna from heaven, they are rich in a variety of nutrients (in addition to iodine), high in fiber and protein, and low in calories. Many species contain higher levels of minerals and vitamins per serving than terrestrial vegetables, including kale and spinach. A plant's roots, after all, only have access to whatever minerals happen to be in their very local bit of soil, and that soil may well be depleted of rarer compounds. On the other hand, seaweeds are surrounded by ocean water full of dissolved minerals, including all those that algae — and we — need in trace amounts. Moreover, because ocean water is constantly mixed by wind and ocean currents, seaweeds always have a fresh supply of minerals to take up.

It surprises me how small a portion of seaweed fulfills a long list of nutritional desiderata. Four sheets of nori — 2.5 grams, or the weight of seven paper clips — delivers a good dose of vitamin A, vitamin B complex, niacin, calcium, magnesium, cobalt, selenium, iodine, and iron, as well as protein (nori is nearly 50 percent protein). Some seaweeds, including nori, contain vitamin C, although that vitamin deteriorates quickly. Nori is also particularly rich in alanine, glutamic acid, and glycine, three amino acids we need to manufacture proteins.

In Japan, people eat on average about 14 grams of seaweed (much of it nori) per day, and they have one of the world's highest average life expectancies. This is not surprising since omega-3 oils are anti-inflammatory and they lower triglyceride levels in the blood, which reduces the risk of cardiovascular disease. The red and green seaweeds produce unique bioactive peptides (small proteins) that have a proven antihypertensive effect. Most seaweeds have a large component of soluble fiber that, like oatmeal, lowers cholesterol levels, maintains bowel health, and contributes to a feeling of satiety. In addition, epidemiological studies show that sufficient DHA intake slows cognitive decline and the progression of Alzheimer's disease. According to the Mayo Clinic, it may help with the symptoms of rheumatoid arthritis. Low in calories, high in nutrients, rich in omega-3s, and filling: better than an apple a day.

Nori "on the hoof" is the red seaweed *Porphyra*, most often *Porphyra yezoensis*. It has long been on the menu in temperate coastal communities around the world, but East Asians, and the Japanese especially, have exploited it as no other people have. *Porphyra* has been such an important part of the Japanese diet for so many years that the population's very biology has changed. The Japanese are better able to digest seaweed because certain of their gut bacteria have genes that direct the manufacture of an enzyme—porphyranase—that helps break down seaweeds' tough cell walls. These gut bacteria likely acquired their beneficial genes via lateral gene transfer from bacteria that successfully inhabit and dine on seaweeds. It may even be that the Japanese are able to squeeze a little bit more nutrition from seaweeds than other people can, thanks to their ability to make porphyranase.

The ancient inhabitants of the Japanese islands were eating seaweeds long before they learned to cultivate terrestrial vegetables. Judging from carbonized remains, coastal tribes were trading seaweed to inland hunting tribes ten thousand years ago. Japanese seaweeds appear often in early written records, in part because seaweeds played a role in Shinto, the indigenous religion dating to the seventh century BC. Seaweeds, including *Porphyra*, were offered at shrines to encourage the gods to protect the sources of this essential food. In the eighth century, fishermen paid their taxes in seaweed, which the tax authority then distributed as a valuable perquisite to court, civilian, and military officials and to priests.

Gathering seaweed to eat, use in barter, or pay taxes became a part of the daily routine of many people who lived along the coast. Seaweeds also became rations for shogunate armies and were eaten as a paste, either sweetened or with vinegar. It wasn't until the early eighteenth century that cooks adapted paper-making techniques to make the now familiar sheets of nori.

Porphyra harvesting has changed over time. Until about 1600, people who lived by the coast simply picked it from the rocks in the tidal zones where it naturally grows. Then, so the story goes, Shogun Ieyasu, head of the military dictatorship that ruled Japan at the time, commanded that he be provided with fresh fish every day at his palace in Edo, as Tokyo was then called. To ensure a steady supply, fishermen around Tokyo Bay constructed bamboo pens offshore to hold fish. It was a happy surprise that *Porphyra* colonized the fences, prompting fishermen to erect fields of bamboo stakes in intertidal waters for the sole purpose of growing the seaweed. Later, in the eighteenth century, Japanese fishermen discovered that *Porphyra* grew not only on stakes, but on nets stretched flat on the water between the stakes, providing a much larger growing surface.

Fishermen came to depend on *Porphyra* for cash income as well as their own food during the winter, but its life cycle was a mystery. In the spring, as the ocean warmed they could see the seaweed released spores and then melted away. In the fall, a new crop would colonize the nets — except when it didn't. In some years and for no discernible reason, the seaweed was a no-show, and the fishermen suffered that winter and cursed the "gambler's grass." From time to time, people would capture some spores and try to seed their nets with them, but these experiments never succeeded. No wonder fishermen offered up a symbolic portion of their crop to the gods each year.

Near the end of World War II and in its immediate aftermath, it seemed that the gods had permanently turned their backs on the fishermen. Again and again, the seaweed failed to appear. Not only was the disappearance of nori a cultural shock, the effects were devastating in a country where people were already starving. Eighty percent of the fishing fleet had been destroyed by American bombs, the food imports on which the country depended had been cut off, and 3.5 million Japanese military and civilians had returned from abroad. *Porphyra* was critical to the survival of fishing communities, yet no one had any idea how to get the seaweed to return.

Salvation came from an unexpected source: a female British phycologist. Kathleen Drew was born in Lancashire in 1901. She not only attended university on a scholarship (unusual enough for a woman at the time), but after graduating with first-class honors in botany, was offered a two-year fellowship to study at the University of California, Berkeley. Upon her return to Manchester, she taught phycology at the university. Following her marriage in 1928 to a fellow academic, Dr. Henry Wright Baker, however, she was required to resign her position because married women were barred from teaching. The university did grant her a research fellowship (unpaid, wouldn't you know), so Drew, as friends called her, quietly continued her work at home. In the next dozen or so years, the slight, bespectacled researcher and mother of two published dozens of papers, earned a doctorate in 1939, and became a leading authority on the red algae group.

In the mid-1940s, she turned her attention to *Porphyra umbilicalis,* a species that grows off the north coast of Wales, where it has long been wild-collected and eaten. Dr. Drew-Baker was determined to solve the mystery of the seaweed's life cycle. To start, she grew the seaweed in small tanks at home in order to capture their spores for further experiments. For no particular reason, she decided to throw some old oyster shells into a few of the tanks. The *Porphyra* prospered and released spores as expected, but a few weeks later, things got strange: the oyster shells turned pink.

At first, it seemed that the water must have been contaminated with the spores of another seaweed. Drew-Baker recognized the roseate fuzz on the shells as a tiny, filamentous species known as *Conchocelis rosea*. But it didn't take long for her to figure out that *Conchocelis rosea* was not a species at all, but instead the sporophyte phase (an intermediate, multicellular phase in the development of certain plants and algae) of *Porphyra*. What she discovered was that *Porphyra* spores don't disappear in the spring, they merely relocate. Instead of recolonizing the tidal area, they sink to lower depths, where they burrow into the shells of oysters and other bivalves and emerge as red filaments. The filaments, which she named *conchocelis* in a nod to their mistaken identity, later release their own spores, called conchospores. Wind and tides carry the conchospores up from the bottom, and they attach to rocks and branches (and nets) in the intertidal zones and develop into the familiar, leaflike stage of the seaweed. The *Porphyra* life cycle is complex, but it has a purpose. The *conchocelis* survive from season to season in the calm depths, providing a reliable source of new life for the species even when storms, unusual heat, or disease decimate the seaweed at the turbulent surface.

Drew-Baker submitted a letter about her findings to *Nature*, which published it in October 1949. No doubt, she expected that only academics with an interest in the red algae group would take any notice. In Japan, however, Professor Sokichi Segawa at Kyushu University read her paper and understood its implications for Japan's nori farmers. The developmental biology of *Porphyra* explained the recent harvest failures. The US Army Air Forces had dropped thousands of underwater mines that exploded in nearly every major port and in the shipping straits. The bombing—intended to force the Emperor to surrender by starving a populace dependent on imported food—also destroyed mollusks and buried the oyster beds. A season of fierce typhoons that followed that disaster continued the disruption of the underwater ecosystem. For years, sporophytes hadn't had a place to embed, grow into *conchocelis,* and release their conchospores to repopulate the tidal areas.

Segawa, with the help of other Japanese marine biologists and fishermen, set themselves the task of creating on land a replica of the ecosystem that leads to the natural reproduction of *Porphyra*. The process they developed varies slightly from country to country today, but essentially it goes as follows: Experts harvest spores from the most productive individual seaweeds in a particular area, then transfer them indoors into large, shallow, concrete tanks filled with seawater. Across the tanks are rods; hanging from the rods by plastic filaments are hundreds of oyster shells, suspended in the water. The water is treated to eliminate bacteria and controlled so that its oxygen, temperature, and nutrient levels mimic local summer conditions. The spores colonize the shells as they would in the wild. After the spores develop into pink *conchocelis,* the water is cooled and agitated to simulate stormy, cooler, fall conditions. At that point, the *conchocelis* release their conchospores. Workers pass fishing nets, wound in layers around what look like miniature Ferris wheels, through the water, and the conchospores attach themselves to the nets. These seeded nets are spooled and frozen until fall, when the fisherman-farmers pick them up and stretch them out in quiet bays. Often, the seeding takes place under the auspices of a local-government-run seeding center.

Within a half-dozen years of the *Nature* publication, Drew-Baker's discovery and subsequent innovations introduced by Japan's scientists had rescued the country's fishing communities and launched a successful state-assisted industry. Gambler's grass is now a sure thing, and *Porphyra* cultivation is a major industry in Japan, Korea, and China, and it has spread to other Southeast Asian countries.

Drew-Baker never visited Japan or knew the impact of her research; she died in 1957 at the age of fifty-six. Japanese fishermen, however, have always been well aware that she preserved their livelihoods and possibly their lives. In Uto, a small town on the Ariake Sea off the coast of Kyushu, there is a monument to the British scientist with her bas-relief likeness in bronze and a summary of her story. Every year on her birthday, April 14, people gather to offer prayers and honor her as the savior of nori. The ceremony is well attended and even today, when memories of the harrowing postwar years are dim or nonexistent, fishing families think of the English scientist with real gratitude. And now whenever I eat my favorite sushi, I give thanks to her, too.

3

On a Grand Scale

Today, sushi has become a common menu item everywhere from upscale restaurants to food courts and college dining halls. People far beyond East Asia have become enamored of the nori-wrapped delicacies (or indelicacies, as traditionalists, dismayed by sushi with bacon, curry, mango, cream cheese, and other unconventional ingredients, might say). In the US, grocery store sales of sushi are increasing at an annual rate of 13 percent and sales of nori snacks are growing at more than 30 percent per year. Consequently, *Porphyra* farming and nori production are now multibillion-dollar businesses. How, I wondered, are fisherman-farmers keeping up with demand? How has a labor-intensive, family-based occupation managed to keep up in the twenty-first century?

In search of answers, I find myself, one February morning, being driven through the countryside of Jeolla province in southwestern South Korea. I am about as far from the capital of Seoul as one can travel in the country, in an area that few foreign tourists visit. With two thousand islands, nearly four thousand miles of coastline, and countless sheltered bays, Jeolla is home to South Korea's seaweed farms.

Today, the sun is brilliant, but the wind is whippy and the temperature well below freezing. The rice paddies on either side of the road are fallow and mostly drained, but remnants of water have frozen into glittering ribbons in the channels between what were once mounded rows of plants. In other fields, miniature by the standards of American heartland agriculture, I see rows of the slender, flaring leaves of spring onions or pale green cabbages the size of basketballs. Farmhouses are scattered across the landscape; their roofs, either bright blue or fiery orange, are gently upswept at their very ends. This traditional architecture contrasts with the utilitarian white hoop houses — nurseries for rice seedlings — alongside most paddies.

Dr. Eun Kyoung Hwang, my host and a senior researcher at the Seaweed Research Center of the National Fisheries and Research Development Institute in Mokpo, South Korea, is driving. She is in her mid-forties with short hair, rimless glasses, and a brilliant smile. Her English is very good, but I sense the effort it takes for her to summon up words in a language she doesn't often use. She and two of her female colleagues are accompanying me on a visit to Aphaedo, one of the largest of the province's islands, to visit a *Porphyra* farm and a nori-processing plant. Dr. Park, who is in the back seat with me, has opened a thermos of a fragrant hot drink, which she explains is made from Asian pear, spring onions, jujube, and ginger. I take a cup while Dr. Lee passes around sweet cookies dusted to a light green with seaweed.

I am not surprised by seaweed cookies; I have been eating seaweed in all forms since I arrived in South Korea a few days ago. My favorite dish, which Dr. Hwang and her colleagues introduced me to at a restaurant the night before, is a soup made with maesaengi, a seaweed that grows in the winter in Jeolla province. I was leery at first: The individual seaweeds looked alarmingly like silky green hairballs and they floated in the clear broth with pillowy, white blobs. But the blobs turned out to be dissolving rice cakes and plump, pale oysters, which I adore. The soup was delicious, with a rich, slightly nutty taste.

Now, as we drive along, Dr. Hwang asks me a question that I suspect she only just feels comfortable asking me, as it implies I may have made a mistake. Why, she puzzles, did I choose to come to South Korea instead of Japan to research nori? Everyone interested in sushi goes to Japan, where the tradition of making it is far older.

Indeed, from the 1960s into the 1980s, if you lived outside Southeast Asia and ordered sushi, you would have eaten Japanese nori. At the time, Japan was the only source of exported nori in the world. In the 1970s, however, South Koreans began to farm *Porphyra* in the wide, shallow waters off their country's southwest coast. Growing conditions were good, and production increased every year. Around 1990, the Chinese entered the business, too, forming joint ventures with the Japanese, who contributed technology while the Chinese provided the labor and coastal locations. By 2013, the Chinese produced 1.14 million tons of nori, the Koreans 405,000 tons, and the Japanese only 316,000 tons.

Japanese production has stagnated in recent years and is unlikely to increase. Rising temperatures in offshore waters are shortening the growing season and encouraging parasites. Industrial pollution has limited the sites where the crop can be safely grown. Also relevant, the life of a seaweed farmer—motoring out in a small boat in the dark early-morning hours of winter to haul wet seaweed out of icy water—holds few attractions for Japan's younger generation, and the country admits few foreign workers who could fill the labor gap.

In contrast, the South Korean government established the Seaweed Research Center and made the development of the industry a priority. Korea now exports ten billion sheets of nori—40 percent of the country's production—each year. Japanese seaweed is of higher quality and is correspondingly more expensive; upscale sushi restaurants buy most of it. (The highest-quality nori is smooth and uniform, dark green to black, sounds like a finger snap when folded, and then seems to melt away in your mouth.) Chinese nori is more likely to be sold into the domestic market or to find its way into fish and animal feed.* South Korean nori, more likely to be dimpled and have some small holes, falls between the two in quality. Because almost all the nori on the shelves of American grocery stores has been harvested from South Korean waters, I tell Dr. Hwang, it seemed most appropriate to visit her country.

---

*Some seaweed snacks in US grocery stores are Chinese in origin, but unless they are USDA-certified as organic, I pass them by. Because seaweeds easily take up and assimilate minerals and metals, they must be grown in clean waters. Coastal water pollution is a serious problem in China, and the regulations on where seaweed can be cultivated are lax.

Besides, she was so enthusiastic when I contacted her, I couldn't resist.

The Shinan Sea Cooperative Association, our destination, has its facilities on the shore of a wide and shallow bay. The tide is hundreds of feet out when we arrive, leaving acres of seaweed nets suspended in midair on skinny poles. Small boats are also stranded; the shining, wet sand mirrors their colorful hulls. When we enter the lobby of the cooperative to meet with Mr. Bae Chang-Nam, we immediately exchange our boots for slippers. Mr. Bae, tall and handsome in a Marlboro-man sort of way, ushers us into his office, where we sit on the sofa and exchange introductions. Tea is offered and accepted. With Dr. Hwang as translator, I learn that Mr. Bae inherited this business from his father, and that his son, the stocky, broad-browed young man who has quietly entered the room, will one day take over from him. Farming *Porphyra* is very much a family and a village affair in Korea. Mr. Bae's operation, though, is unusual in that it also processes its crop. Most fishermen-farmers sell their seaweed to processors at auction, but Mr. Bae has a more entrepreneurial bent. By controlling the downstream processes, he captures the profits from all stages of nori production.

Dr. Hwang and her colleagues speak in Korean with Mr. Bae, and the light tone of pleasantries gives way to a more serious conversation. Then our host stands up, smiles, and it's off with the slippers and back into boots. We head into the cold to trace the voyage of *Porphyra* into sheets of nori.

The first stop on the tour is an aluminum shed that shelters a half-dozen large vats. In each vat is a dark suspension, a mixture of freshly harvested *Porphyra* blades, which look like scraps of satin ribbons, and water. A giant, rotating daddy longlegs of a machine sits on top of the vats, stirring the contents. After several rinses to remove sand, parasites, and epiphytic algae, the blades go through a mincing machine, where they become a dense, nearly black slurry.

The slurry is then channeled indoors, where it flows into an industrial nori-making machine, a clanking, whirring, thumping blue monster dozens of feet long. Running through the heart of the monster is a sixteen-foot-wide gray conveyor belt that is striped with rows of yellow mats. At the far left of the belt, spigots spray neat black squares of *Porphyra* slurry onto a row of mats so only a frame of yellow is visible around each. Every two seconds, as the rows proceed down the length of the machine, steel beams running perpendicular to the belt briefly thump down, pressing the water out of the patches of slurry. At the far end of the room, the belt enters a blue industrial oven the size and shape of a storage pod, where it folds like an accordion and pauses as the slurry squares are thoroughly dried. Then, the square sheets of nori are flipped off their mats and onto another belt that carries them through a series of other rooms, where they are toasted, inspected, bundled, wrapped, and boxed for shipment. From slippery seaweed to elegant packages of tissue-paper-thin nori takes only a few hours.

On the way back to Mokpo, Dr. Hwang explains why she has been meeting with Mr. Bae. A community seaweed research group has been offering farmers in Mr. Bae's area a new variety of *Porphyra* it discovered. At first, the new variety seemed like a winner, developing bigger and more robust blades and significantly more mass. But the texture and the taste of the nori, Mr. Bae decided, are inferior. Although he has chosen not to culture it, his neighbors do. Now he finds that the new variety is aggressive. As soon as he puts his seeded nets into the water, the new seaweed colonizes them and outcompetes the traditional variety he has built his company on.

It is a difficult problem, Dr. Hwang says. She needs to tread carefully with the local group to make sure she doesn't disparage its efforts or appear highhanded. Mr. Bae's dilemma highlights one of the peculiarities of sea farming: farmers have little control over their "soil." In that way, sea farming is more like English farming in the days of the commons. Nonetheless, Dr. Hwang's research group is exploring solutions. She hopes that if Mr. Bae sets out his nets earlier in the season, his *Porphyra* variety will be sufficiently established to fight off the introduced one, but there may be other ways he can change his cultivation methods to control his crop.

The next morning, after my breakfast of seaweed-and-turnip soup with rice at the hotel, Dr. Hwang picks me up, and we head south across the Myeongnyang Strait to Jindo Island. The two bridges that cross the wide strait are dominated by sets of triangular, fire-orange towers that look ready to launch themselves into the deep blue sky, restrained only by glinting silvery cables that stretch from the towers to the bridge deck. Ninety minutes beyond the bridge is our day's destination, Hoedong Harbor. I am already bundled from chin to knees in a down coat and from knees to toes in waterproof boots, but when we walk out onto the enormous, vacant, concrete quay, the wind stings, and I add a hat and gloves. We board a boat waiting at the end of the quay, and the captain backs out into the harbor.

There is no romance in going to sea on a seaweed boat. The deck of the long, twenty-foot-wide boat is covered almost completely from end to end and side to side with a three-foot-deep blue bin. Captain Lee steers from what looks like an orange-and-blue phone booth at the right side of the stern. Alongside the phone booth and spanning the width of the boat is a horizontal cylinder with multiple blades; it looks like the business end of an old-fashioned push reel lawn mower, only on a gigantic scale.

As we motor out of the harbor and into open water, we have a 180-degree view to the horizon; ahead of us, the entire expanse is covered in floating nets. Ten minutes later, Captain Lee slows the boat as we approach the edge of this vast array, and I see that the nets — each long and narrow — are organized like rows of crops on a terrestrial farm. The rows are separated by slender alleys just wide enough for our boat. A "plot" — that is, one farmer's set of nets — has a wider lane around it, demarcating it from its neighbors. All the rows and all the separate plots are perfectly symmetrical and absolutely identical. There are no markers anywhere, and I can't imagine how any individual farmer locates his particular plot. But not only do farmers know which plots are theirs, they know the plots' varying levels of productivity. No family owns its plot. Every year the farmers (who are organized in a cooperative that leases the area from the government) hold a lottery to select plots for the coming season. Everyone knows which plots have the best currents – neither too fast nor too slow – and are best sheltered from wind and waves, and chooses accordingly.

The nets are about one hundred feet long and eight feet wide, and they're anchored at either end to the ocean bottom – too deep for us to see – to keep them in place and stretched out. They float, thanks to what look like Styrofoam barbells that are attached along the length of the next every dozen feet or so. Some nets lie on the water's surface with the barbells on top of them; others lie across the top of the barbells and thus are elevated about eighteen inches above the water. Hanging from the elevated nets are limp, wet streamers of *Porphyra.*

In the wild, *Porphyra* growing on rocks are exposed to the air for a few hours twice a day when the tide goes out. The seaweeds depend on that exposure: parasitic snails and minuscule crustaceans that nibble the blades (and microalgae that block sunlight) can't abide being dried out, so they either drop off or die. When farmers hang their nets on stakes in tidal zones, they attach the nets in a way so that, when the tide retreats, nets are temporarily suspended above the water's surface. But in these deeper waters, the farmers mimic that air drying by attaching the nets to these Styrofoam barbells. Four times a day, the farmers motor out to turn over the nets.

This sounds terribly labor intensive, but Dr. Hwang assures me it's not, and she asks the captain to show me how it's done. He idles the boat by the end of a net, and one of the two crew simply grabs the endmost barbell and flips it 180 degrees, which causes all the other barbells to turn over, one after another, like dominoes falling. The switch takes less than a minute.

We pull up to another boat, which is in the process of harvesting seaweed from a net. The crew is pulling the net over the rotating blade, shearing the seaweed from its underside. The trimmed ends fall into the blue bin, and the net continues over the boat, off the front end, and back into the water. During the height of the season, the seaweed gets a shave every two weeks. As the weather warms in March, the seaweed's growth rate gradually slows and the fishermen retrieve their nets, just as the fishing season starts.

We return to the harbor at about 10:15 a.m. and find the deserted quayside transformed. More than two dozen seaweed boats cram the area, with a row of boats tied to the quay, others lashed to them, and then others to those, all in a disordered scrum. The scene is a crazy quilt of color: the boats are blue and orange, the mooring bumpers are bright yellow. The farmers—mostly men but some women, too—are dressed in rubber boots, waterproof overalls, puffy jackets, and hats. Every article of clothing is a different bright color, a vibrant medley of red, blue, purple, yellow, lime green, and pink. This is serious business, but everyone is dressed as if they were going to kindergarten.

The bins on the boat decks are full of wet *Porphyra* that, in such bulk, looks like black sludge. All the boats arrived at the quay at the same time; everyone waits exactly 45 minutes for the auction to begin at 11 a.m. The waiting period allows everyone's haul to drain, so that the buyers, who pay by weight, are buying similar quantities of seaweed.

At 11 a.m. sharp, a man in a red baseball cap steps onto one of the outermost boats, reaches into a hillock of seaweed, and begins the grading process. He is accompanied by a dozen potential buyers who bid (or don't) on the crop of each boat. It doesn't take long for the auctioneer to make the rounds, stepping from boat to boat, and the entire morning's haul is sold. Then, workers on the boats, knee deep in cold, wet seaweed, start shoveling their harvest into blue or green polypropylene bags five feet high and just as wide. As the bags fill and are drawn closed, behemoth dump trucks rumble onto the quay. Towering cranes swing over the boats and in a single pass grasp a bunch of bags by their drawstrings, hoist them into the air, and then drop them into the trucks. By noon, the concrete quay is empty again and quiet, but only until the next morning. South Korea's nori industry has been growing at 8 percent per year, and these quays, like many others all along the southern coast, are getting busier.

The South Korean government understands the economic potential of farmed seaweed, which is why it has invested in Dr. Hwang's research center. The waters off the southern coast are clean, and there are hundreds of miles of deeply indented coastline that could be productive. Sea vegetables are the vegetables of the future.

4

Welshmen's Delight

Porphyra yezoensis grows only in East Asia, but a number of other *Porphyra* species grow in waters off Europe and North America. For hundreds of years, the Welsh have been wild-harvesting *Porphyra umbilicalis*, commonly known as laver, from coastal rocks. In 1607, William Camden wrote in his encyclopedic *Britannica* that the peasantry along the Pembrokeshire coast gathered a "kind of Alga or seaweed" that they washed of sand, dehydrated between two tile stones, shredded and kneaded, and then boiled for many hours until it became a glutinous deep-green mass called laverbread. (It's the kneading that led to the confusing name of a food that is pure seaweed.)

Laverbread has traditionally been eaten in the morning; it can be spread on toast (for the purist), mixed with porridge, or shaped into patties and fried in bacon fat. It has long been on the menu for dinner, too, in a creamy stew with cockles, small clam-like mollusks with a sweet and briny taste. Laver has been a staple of the Welsh diet for centuries, and it's a good thing too: a serving provides as much as 6 grams of protein, 33 percent of vitamin A requirements, 50 percent of vitamin C needs (if eaten promptly), a good dose of B vitamins, and other minerals, including iodine and iron.

In recent decades, though, the Welsh have drifted away from laver dishes, and most Welshmen today rarely, if ever, eat it. Jonathan Williams, the founder and owner of Pembrokeshire Beach Food Company and the proprietor of a mobile eatery on the southwest coast of Wales, is out to change that. I became aware of Jonathan when I googled "Wales seaweed" and found that nearly all the results related to his businesses. So, on a warm and rather rare sunny day in mid-June, I head to Freshwater West beach in the Pembrokeshire Coast National Park where Jonathan parks his Café Môr, the food trailer he started seven years ago.

There are two parking lots at the beach, and I have inadvertently parked in the farther one, which is a couple hundred feet from the shore and on top of a cliff. I hike across a meadow dotted with white and yellow wildflowers and look down. Below me in either direction is a gentle two-mile smile of a beach whose sandy sweep is broken up by the occasional rocky outcropping. The tide is out, way out — I'm guessing it's five hundred feet from the cliff to the water. The sea is calm today, and sends long lines of well-mannered waves to curl and break on the shore. I'm surprised to see wetsuited surfers, but in fact I'm looking at one of the best surfing beaches in the UK, and today is perfect for novices. Despite the water's placid appearance, I know that rocky reefs beyond the breakers have wrecked many hundreds of sailing ships over the years. In a single storm in 1703, thirty ships foundered here.

I make my way down the cliff and walk a half mile along the water's edge and back up the cliff again to the other parking lot. I had worried that I might not find Café Môr, but now I see that no one could overlook it. Not only is it the only eatery at the beach, but the boxy trailer (what the British call a caravan) is painted bright blue, and its side panels are flung open to display menu boards exuberantly lettered in yellow, pink, and aqua chalk and embellished with beaming suns, spouting whales, and stylized waves. Solar panels cover the roof. Jonathan — tall, dark-haired, bearded, and at the moment very busy — is one of two cooks behind the counter, his head nearly grazing the caravan's ceiling. I know he won't be available to talk until 2 p.m., after the midday rush, but I've come early to make sure I have time for lunch. I get in the queue (a queue that didn't disappear until the 4 p.m. closing) and order a sandwich of Pembrokeshire lobster in Welsh Sea black butter. Black butter, it turns out, is laverbread cooked in butter and lightly spiced with paprika and black pepper. I take my sandwich across a narrow access road and perch on a cliff that overlooks a scrap of beach where a family is playing bocce and a girl is trying to fly a red box kite.

The sandwich is deeply, surprisingly flavorful. I know the tastes of lobster, butter, paprika, and pepper, but they don't add up to this richness. It must be the inclusion of laver. I can't put my finger on a flavor – there is no seaweed taste per se, but the laver seems to deepen the flavors of the dish, like adding a bass to a string trio.

After lunch, while waiting for Jonathan, I walk along the cliff where the meadow is springy underfoot. I watch as dozens of little sand martins swoop across the blue sky and then, like swiftly returning boomerangs, vanish into their burrows in the cliffside. Farther on, I discover a simple hut, its two wooden sides leaning into each other in an A-frame. A placard informs me that laver collectors once draped their seaweed over the hut's roof to dry, and that this is the last remaining of twenty such structures that once dotted the Freshwater West cliffs. The huts belonged to families from the nearby town of Angle who supplemented their incomes by selling dried laver to the factories in Swansea, sixty-five miles down the coast.

The laver-collecting business began in Freshwater West in 1879, after an American merchant ship on its way to Liverpool ran aground on the reefs in a severe gale. Local rescuers shot a lifeline from shore to ship and saved all but two sailors. Meanwhile, 15,000 boxes of cargo (including 214 sacks of valuable mother-of-pearl boxes) were strewn over the reef and beach. Word of the bounty spread quickly, and scavengers arrived from as far as Swansea. It was the Swansea folks who noticed there were more than temporary treasures to be had; while wading in the surf, they recognized thick beds of laver. They subsequently contracted with the families of Angle to collect and dry it, then take it by horse-drawn cart to the Pembroke train station, from whence it traveled to Swansea. The factories in town processed it into laverbread and sold it along with cockles plucked from the rich beds at nearby Burry Inlet. The Freshwater West cottage industry provided seasonal income for the local residents until the 1930s, when the factories sent their men by truck to gather the laver themselves.

Jonathan, having doffed his apron, joins me at the hut. "I grew up in Pembroke, about ten minutes from here," he tells me, "and as kids we used to come all the time with my parents. Later, I learned to surf, and we used to have beach parties down here. The hut was a reminder of a way that once was, although," he says, laughing, "for us, it was mostly a good place for making out."

After university in Swansea and several years as a consultant, Jonathan changed course. He was thirty years old, he says, and sitting at his computer at work one day, when he realized how deeply he missed the sea, Pembrokeshire life, and – having grown up in a family that loves to cook – the food. After some serious thought, he told his boss he needed to work part-time so he could start a food business that would serve the seafood and sea plants he knew from his childhood. In the summer of 2010, he set up a Saturday food stall outside a farm shop near Pembroke. "After the first day, I realized I had met so many people and talked so much that my jaw ached. I also realized, with the hours of preparation, I'd ended up working for less than two pounds an hour. Still, it felt like the best day's work I'd ever done." Jonathan had found his calling.

Seven years later, he has three closely related operations, including Café Môr (môr means "sea" in Welsh), whose business is booming. On weekends, he often has five cooks in the caravan, standing literally shoulder to shoulder. (He prefers the term *cook* to *chef*: "Chefs are generally miserable down here; they should be in kitchens, not out on display.") He also has two other mobile food trailers – a pop-up shack and a fishing boat on wheels – that travel to more than thirty festivals around the UK each year.

The third business is a processing and packaging facility in the nearby town of Pembroke Dock that turns out seaweed products that are sold in 400 outlets around the UK and increasingly overseas, too. Pembrokeshire Beach Food's bestsellers include Welshman's Caviar, which is Welsh laver dried, shredded, and toasted; Ship's Biscuits, which are fish-shaped crackers baked with laver and topped with dulse (another red seaweed) and sea salt; and Welsh Sea black butter, the laverbread sauce that enhanced my sandwich. The company also sells packaged dulse, kelp, and other seaweeds, locally harvested or brought in from Scotland and Ireland. "Demand for seaweed has gone crazy," Jonathan says. "At a trade show in London recently, I had serious inquiries from Heinz and other ready-food manufacturers, looking to up the nutritional value of their products."

It's too early for large-volume contracts, though. The problem, Jonathan notes, is that "laver is a mysterious seaweed." Like its wild East Asian counterparts, it sometimes fails to put in an appearance or arrives only in small numbers. In Wales, winter storms batter the coastline, rip all the seaweeds off the rocks, and may derail conchospore production in deeper waters. Recently, Jonathan contracted with Dr. Jessica Knoop, a marine biologist at Swansea University, to conduct research on the life cycle of laver, with the hope that he might someday cultivate the seaweed the way nori is cultivated in East Asia. But until then, he can't guarantee the steady supply that major processed-food companies would need.

This may be a good thing. Gratifying as the growing interest in Welsh laver is, Jonathan is sensitive to the potential impact of too much laver love. "There's a danger of food companies wanting to provide a quick nutritional fix for their products, and come and scrape up everything off the bottom. For now, it's just me and the Swansea boys, and they're not here every day." (The "Swansea boys" are Selwyn's, Ltd., and Penclawdd Shellfish Processing, Ltd., which process and sell cockles and mussels, as well as laver and other seaweeds.) "One of the problems is you'd never be able to see what damage is being done. It's not like the rainforest, where you can see what's being cut down." So far, at least, it's just apprehension, not reality.

We hike down to the beach, then into the shallows where sand gives way to rocky reef. Jonathan leaps from one outcropping to the next while I pick my way more carefully, but we're both soon up to our knees in cold water. (Even at the height of summer, the water temperature doesn't get warmer than 55 degrees Fahrenheit or about 13 degrees Celsius.) There's plenty of gutweed, *Ulva intestinalis*, which looks like handfuls of carrot peelings, only bright green. "Poor man's laver," Jonathan says. "Not much flavor in *Ulva*." Next, he stops to pinch off some seaweed that grows in little chocolate-colored, hand-shaped fronds that cling tightly to a vertical rock face. "Pepper dulse, on the other hand, is lovely. We put it in salads and cook it in pasta." I nibble it. Indeed, it's peppery and delicious, like a crunchy arugula. In the many tidal rock pools, Jonathan points out other species—rusty or green or golden, stringy or fernlike—clinging to the rocks, waving gently in the glass-clear water.

The laver peaked this year in May, and there's not much left now, but Jonathan, sprightly and splashing here and there while I clamber in pursuit, finally spots some and pulls it off a rock. Stretched out between my hands, it's nearly transparent, astonishingly thin, and curiously stretchy, like a piece of green plastic food wrap. There are some eight hundred kinds of seaweed off the coast of Wales, he tells me, and people have only tasted about fifty species. That fact clearly intrigues him, and I can see he imagines all sorts of untapped culinary possibilities. It makes me wonder: What if we knew only 6 percent of the vegetables we now eat? Think what we would be missing. It's likely, for example, that the thorny, thistly, leather-leaved cardoon—a weed that farmers selectively bred until it became my favorite, the globe artichoke—would not have been among that 6 percent.

Which reminds me that many familiar vegetables wouldn't be on our plates if we hadn't selectively bred them for thousands of years, gradually altering them to become more appetizing and tender. The cardoon; the tiny, tough teosinte that is ancestor to maize; and the brown root that became the carrot were all once barely edible. Sea vegetables can be as tasty or tastier than terrestrial vegetables—I'll take pepper dulse over iceberg lettuce any day—but even those less toothsome varieties might be enhanced if we work on them. We've cultivated only a few sea vegetable species. Who knows what culinary treasures are out there?

It may be harder to select and breed vegetables that live in wild water instead of cultivated soil, but we've already been doing it. When Jonathan traveled to Japan last year with a British trade delegation, he met eighty-year-old Dr. Fumi-ichi Yamamoto, who has been involved in *Porphyra* farming since the 1950s. When Dr. Yamamoto tasted some Welsh laver that Jonathan had brought with him, he said it reminded him of the rich taste of the wild nori of his boyhood. Like modern tomatoes, Japanese *Porphyra* has been selectively cultivated to survive the conditions of modern cultivation, processing, and shipping. (In South Korea, Mr. Bae's neighbors are similarly selecting for characteristics other than traditional taste.) But there's no reason we can't discover and breed particularly tasty seaweed varieties.

Jonathan has started a revival of traditional seaweed dishes in Wales, but whether *Porphyra umbilicalis* will become a significant crop there is an open question. It would be gratifying to think that Kathleen Drew-Baker's work, which was so instrumental in building the East Asian industry, could create a British one, as well. However, what Jonathan is doing for seaweed cuisine is also significant. Cooking seaweeds in Asian dishes is wonderful, but we could use more recipes for incorporating seaweeds, including nori or laver (which are interchangeable in recipes), in the regional dishes of other cultures. Pembrokeshire Beach Food has recipes on its website and adds new ones frequently. In a world where protein and nutrients, arable land, and fresh water are increasingly in short supply, adding sea vegetables to our diets will suit us all. Bravo to the chefs (and cooks) who love seaweed for its taste, as well as its nutrition.

5

A Way of Life

When it comes to edible seaweeds, *Porphyra* is the most familiar, but if you're a fan of miso soup, you've also gotten up close and personal with the seaweed known as wakame (*Undaria pinnatifida*). I have always enjoyed miso soup, yet until a few years ago, I would drink the broth and eat the tofu, but navigate around the green bits, leaving them at the bottom of the lacquered bowl. Unlike the dried nori in sushi, floating wakame actually looks like seaweed.

It was a serious sin of omission, I now know. For one, wakame is very tender and slightly sweet. Miso soup is a delicate mix of tastes, and wakame adds some expressive notes to the composition. It is also nutritious, particularly rich in iron, calcium, and magnesium, and has generous amounts of vitamin K and folate, an important B vitamin, as well.

It is not only the wakame, though, that is important to the harmony of the soup. Dried kombu, a term for various species of the brown seaweed genus *Laminaria,* also plays a role. A four-inch piece of kombu added during simmering and removed before serving consecrates miso soup—or the stock of any soup or stew—with the savoriness known as umami. Umami is one of the five basic tastes, in addition to salty, sweet, acidic, and bitter, that we perceive via our taste buds alone. (We actually smell other tastes via receptors in our nasal passages that trap molecules rising from food.) We taste sweetness only at the front of the tongue, bitter at the back, and sour and salty at the sides, but the taste buds that perceive umami are distributed over the tongue's entire surface.

Umami was discovered in 1908 by the chemist Kikunae Ikeda of Tokyo Imperial University when he was trying to understand why kombu imparted such flavorfulness to miso stock. Its source, he discovered, is glutamate, a salt of the amino acid glutamic acid. One serving of wakame has a small amount (2 mg–50 mg) of glutamate, but a serving of kombu can have as much as 3,000 mg. Later in the century, Japanese chemists recognized that miso and bonito fish flakes contain other amino acid salts that also confer an umami taste. When two or more umami-containing foods are mixed together, such as the wakame, kombu, miso paste, and bonito flakes in miso soup (or the tomatoes, parmesan, and anchovies on pizza), the combined flavor intensity is greater than the sum of the parts. Given its ingredients, it is no surprise that miso soup is one of the most frequently consumed foods in Japan. Seventy-five percent of Japanese people eat the soup at least once a day; more than 40 percent eat it twice a day.

Not only do Japanese people tend to live long lives, but Japanese women have among the lowest occurrences of breast cancer in the world. In the US, one out of eight women will be diagnosed with breast cancer in her lifetime, but only one out of thirty-eight Japanese women will receive this frightening diagnosis. Cancer researchers agree that the traditional Japanese diet plays a major role in the low incidence of the disease. The mechanisms of diet and breast cancer are complex, and one's genes certainly play an important role. Nonetheless, a 2003 study by the prestigious *Journal of the National Cancer Institute* concludes that "frequent miso soup and isoflavone [a soy compound] consumption was associated with a reduced risk of breast cancer." Other studies report a benefit from seaweed alone, but there isn't enough research to consider the question settled. I eat miso soup for its unmatchable savoriness, but if it has an anticancer benefit, all the better.

Wakame, kombu, and other brown seaweeds — which people often group together and call kelp — grow in all the world's oceans. Latecomers to the algae party, they evolved between five and twenty-five million years ago into a range of forms, from tiny filaments to the giant kelp that grow as forests off the coast of California. By mass, the Phaeophyceae are the most dominant group of seaweeds in the ocean. Most browns live in colder latitudes, but free-floating *Sargassum* are happy to hang out in warmer climes. Most of its species are long, rubbery, branching, and buoyed by pneumatocysts. You can find them most abundantly in the three-million-square-mile Sargasso Sea, a region of the temperate North Atlantic where, entangled with each other, they form drifting, gold-toned islands. The islands are oases for sea turtles, eels, snails, crabs, juvenile fish, and other marine animals that live within or on them. Many of these creatures – some of whom live nowhere else – have evolved their own golden hues, camouflaging themselves to hide from predators.

Some East Asians eat *Sargassum* boiled or fried and stuffed in dumplings, but milder – and more popular – browns are found in colder waters off the coasts of Asia, Europe, and North America. Which is why, on a July afternoon, I am perched on a wooden box at the squared-off bow of a wooden boat, dressed in a wetsuit and neoprene booties. Larch Hanson, the owner of Maine Seaweed Company in Steuben, Maine, is piloting the boat, and we're on our way to Gouldsboro Bay, where Larch, a guru among wild-seaweed harvesters, collects kelp. This open boat, a traditional fisherman's garvey, is sixteen feet long and black from decades of linseed oil applications. A nine-horsepower Johnson engine just inside the transom powers the craft. Trailing us, attached by an eight-foot line, is another garvey and then, attached by two more lines, is a pair of rowboats – miniature, snub-nosed, snub-tailed versions of the garvey – that bob along behind us like ducklings.

It is noon on a cloudless July day, but I have a down jacket and a wool cap stuffed in my lap, gear that Larch insisted I take from a rack in his house. It's not clear to me why he also insisted I wear a wetsuit since I'm not planning to get in the water, much less why, on a warm summer day, I need winter gear. The booties, though, have already proved helpful. Larch goes out to collect seaweed at low tide, which means we had to squelch across a boggy beach, then wade through the water to reach his little fleet anchored offshore.

I had driven up from Boston that morning and arrived at Larch's house when it was already an hour past low tide. He was anxious to get going, so I hadn't had much time to talk with him before we set out. Now, I'm full of questions, and I turn around to ask them.

He stands in the back of the boat, tall and slender and fit in his wetsuit, looking like a high school distance runner. His posture is perfect; his complexion fair and fresh. His hair is fine and straight and snow-white, cut in a style that you might see on a ten-year-old boy in a Norman Rockwell painting. I know he is sixty-nine years old, but he looks much younger. I call out a question to him, but I have a soft voice, and even though I'm shouting, it's hard to make myself heard over the engine. "Wait until we get there," he shouts back. So I hold my thoughts and simply enjoy the trip up the coast, watching the flat, deeply forested coastline and the seaweed-covered beaches scroll by. Seabirds slice the sky in the distance or, closer, swoop in to land on rock outcroppings.

After about thirty minutes, we pass a point of land and the wind picks up and the water gets choppy. I zip up my wetsuit to the neck and put on my jacket. Thirty minutes later, just as Larch cuts the engine and hooks a mooring buoy, I gratefully add the hat. We are about a quarter-mile offshore. All around us in the water, just under the surface, I see large, shadowy masses of brownish seaweed. What I'm looking at, Larch explains in a mellow, Midwest-accented voice, are the blades of sugar kelp, *Saccharina latissima*. The kelp are growing from the bottom, where their holdfasts secure them to a submerged ledge. A long stipe, which I cannot see, leads from the holdfast to a single, long, substantial blade. Sugar kelp grows in cold waters in areas where the tides are strong enough to bring fresh supplies of nutrients, yet sheltered enough so waves and storms don't tear the holdfasts from their anchorage.

All around the boat, blades of kelp wave and swirl just under the surface. They are essentially weightless in the water, buoyed by their hollow stipes. At high tide, only their tips are at the surface, but at low tide, right now, the blades float horizontally and are easy to see. Larch leans over the side and, reaching down deep into the water with both arms, pulls up a mass of kelp blades and stipes. With a quick slash of his serrated knife, he hauls the whole lot into the boat. He isolates a single blade from the mass to show me; it looks like a ten-foot-long, one-foot-wide lasagna noodle, only golden-brown. I run my hand down its flat center and up a ruffled edge. The ruffles create turbulence, Larch explains, stirring up the

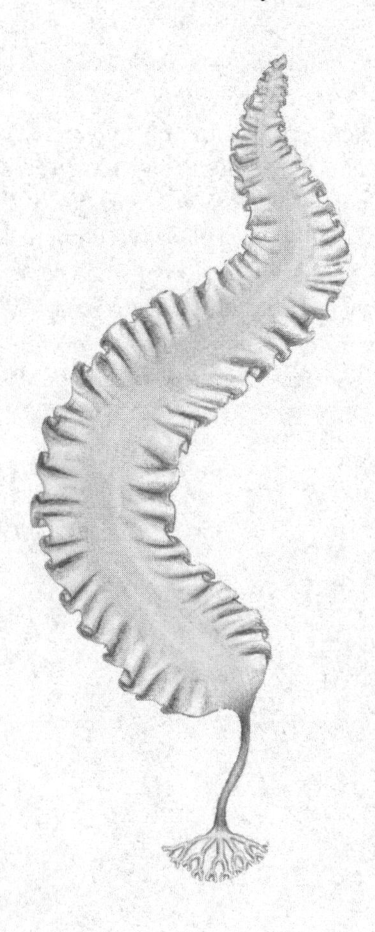

Saccharina latissima

water and bringing more nutrients to the blades' cells. The noodle feels *al dente,* and it's as slick as if it had been tossed in cooking oil. I lift one end up toward the sun: it is heavy but translucent, and, with the sun behind it, it glows like a golden-green pane in a stained glass window.

I ask about the holes, some as big as pennies, that pock the uncut end.

"Ah," Larch says, "it's late in the season for kelp. The blade grows from its lower end, just above the stipe, so you're looking at the older end, and it's deteriorating. Snails and other parasites have been eating it." He flicks off a snail the size and color of a carob chip. "This piece still has good color, nice and dark, but we'll put it through the mill and chop it up for our soup mix."

The season for collecting kelp is short. In May and June, Larch and his apprentices go out regularly, always at low tide. In addition to *Saccharina,* they take other brown seaweeds, primarily *Laminaria digitata,* whose sheaf of long, skinny blades resemble a hand, and *Alaria esculenta,* whose blades have pairs of small reproductive structures like dragonfly wings at their base. Alaria, which has a slightly nutty taste, is often used as a substitute for wakame, and *Laminaria digitata* and *Saccharina latissima* can substitute for Japanese kombu in recipes.

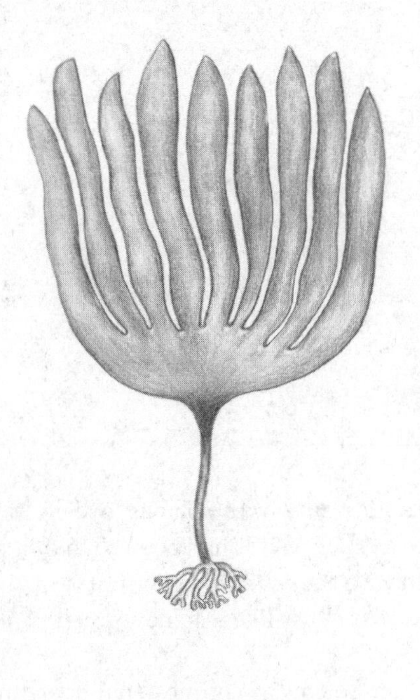

Laminaria digitata

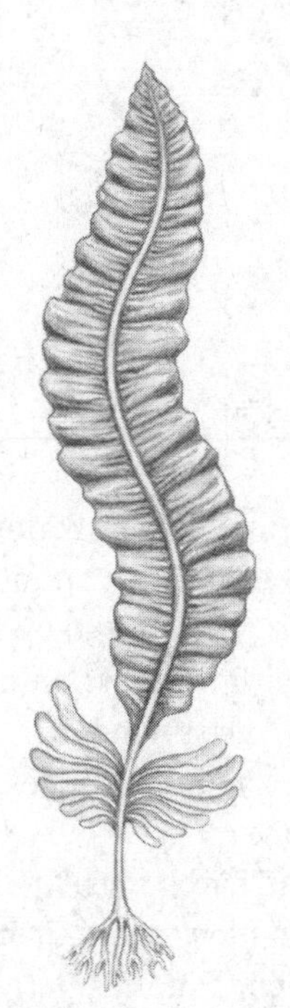

Alaria esculenta

Alaria lives at two depths. Its perennial beds lie deep enough that they aren't wiped out by ice storms. The shallower annual beds are like a blackboard that is erased every winter. Larch doesn't harvest the perennials, leaving them to spawn and produce spores that regenerate the annual crop. The *Saccharina* species and the *digitata* live for two to four years and, if their life cycle isn't complete, their blades will regenerate after he trims them. They persist through summer, but their growth slows and they become paler as the water warms. He continues to harvest them, but less frequently, and some will end up as compost for his garden.

Larch has been harvesting for four decades. "In my brain," he says, "there's a map of the islands. I can see how many bushels of *Alaria* I'll get from each ledge, and how many boatloads of *Laminaria* I'll get. All these kelps grow in the same beds that were marked in the tax maps in the assessor's office back in the 1960s. In other words, I haven't taken too much of anything." Here, under his care, the kelp is a wild but managed crop.

And now, Larch is ready to start harvesting. He pulls one of the smallest rowboats, about as long as he is tall, to the side of the garvey, steps in gracefully, and, seated facing the stern, rows a dozen yards away. Only about eighteen inches of freeboard separate him from the water. Crouching in the bottom of the little boat, he plunges his arms into a mass of kelp and immediately falls into a rhythm: Reach, grab, pull, slice, haul; reach, grab, pull, slice, haul. When his boat is full, he rows over to the unoccupied garvey and transfers armfuls of kelp into it. Then he rows back and repeats the process. This harvesting would be hard enough if the sea were a placid pond, but it's not. The rowboat rises and falls erratically. I can see why, with a smile, he declined my offer of help before we set out. Unsettled sea, sharp knife, slippery seaweed, novice collector. Not a promising mix.

Two hours and many loads later, we motor home. Larch, unaffected by such a vigorous and surely back-straining workout, makes a small detour to a nearby island, where we watch dozens of seals playing in the water or sleeping on the beach. The garvey is heaped full of kelp, a glistening mass of autumnal tones in the afternoon light, a boat-shaped bowl of lasagna ready for a giant's dinner.

The tide is higher when we return, but we still have to moor the boats in shallow water and wade back to shore. Later in the evening, when the tide is fully in, his apprentices will row the garvey up to the top of the beach, unload the wet kelp into bushel baskets, dump them into open-top trailers attached to a small tractor, and drive the trailers to the drying shed. But for now, Larch and I head down a path in the woods and across the road to his home. We pass four blue tents on wooden platforms that are currently occupied by apprentices. They have stayed on past the prime season, both to help in the last days of harvesting and packaging, and also to build a new drying shed. Another tent is taken by a young family; a sixth by a customer who's been buying seaweed for decades and makes an annual retreat here; and a seventh is occupied by a woman who runs a yoga center where Larch will teach in the fall. At the next tent, in a grove of conifers, he peers inside and deems it uninhabited, and we perch on the edge of the platform. The sun filters through the branches, the air is warm and smells of resin and ocean, and a bird trills a two-note song.

Larch has been collecting kelp by boat from areas where he alone has permission to harvest. He also scavenges dulse, collecting it from beach rocks. In a typical year, he'll sell about 6,000 pounds of dried seaweed, which means he and his apprentices will have pulled 60,000 pounds of wet seaweed from the ocean. Most of his dried products — *Saccharina, digitata, Alaria,* and a little dulse — he ships to individual customers who buy it in three-pound bags and cook with it. Recently, he has been selling about 1,200 dried pounds to a cosmetics company. Another 400 pounds or so is rockweed, a species that grows densely along the shore, that goes to a company selling iodine tinctures. Since the Fukushima nuclear power plant disaster in 2011, business from West Coast buyers has doubled. Some people believe that by absorbing iodine from Maine kelp, their bodies will not absorb radioactive iodine from fish and other sources contaminated by Fukushima radiation.*

---

*In theory, this is true. However, a 2013 study supported by the National Oceanic

Larch is not an ambitious man. He has all the business he needs or wants. He proudly says he has always operated on a cash-and-carry basis, lived by his labor, and provided for his family, which includes four children. His heart is in developing a community of wild seaweed harvesters, men and women who responsibly tend the great underwater vegetable garden off the coast of Maine. Each spring and summer about half a dozen young people apprentice with him to learn his methods. Some return repeatedly. The work is physically demanding: those sixty thousand pounds get lifted six times — from ocean to rowboat to garvey to bushel baskets to trailers to drying racks — in ten weeks. The work hours are odd: hitting the low tides sometimes means going out in boats in the middle of the night. The pay is a stipend, plus a place in a tent and a seat at the communal table.

Wild seaweed harvesting hasn't made Larch a fortune, but it is a way to make an independent and healthy living. Several of his apprentices have set up shop themselves, which not only gives him the satisfaction of seeing his way of life spread, but has a practical value, too. Demand for and supply of seaweed species varies, and from time to time he trades with fellow collectors to fill orders. No money changes hands, he emphasizes; it's always done "pound for pound." I ask him what he thinks about a few small companies that have started farming kelp, and am surprised, for a man whose middle name might be Equanimity, by the vehemence of his feelings. "They grow their product too intensively," he insists, so the crop is less nutritious and maybe more prone to pest damage. Moreover, these farmers, "who are potentially my competitors," receive state government support ("my tax dollars") for research and development of aquaculture methods, which outrages him. Later, I read this on his website: "What are the qualities of a plant that thrives in surf? Tenacity. Flexibility. Joy in the moment. Watch a crew harvesting *Alaria*. They are developing these same traits. It's no

---

and Atmospheric Administration concluded that "additional doses from Fukushima radionuclides . . . are comparable to, or less than, the dose all humans routinely obtain from naturally occurring radionuclides in many food items, medical treatments, air travel, or other background sources."

wonder that I don't ever want to see *Alaria* become 'domesticated' via aquaculture."

We head back to the house. On the first floor is the company office. A taxidermied fish hangs on a wall; this Northern pike was Larch's first catch as a child growing up in rural Minnesota. The walls throughout the house are unfinished pine board, and their scent is in the air. I return my borrowed jacket to hang on a rack full of others — a supply for visitors of all sizes — and follow Larch up a steep staircase to the third floor. Here, under the roof, most of the space is taken up by a common room. A long wooden kitchen counter with open shelving above stretches the length of one side. Its surface is covered with dishes, cookware, containers of beans and spices. A large wood-burning stove stands by the door.

Larch introduces me to his wife, Nina, who is chatting with several twenty-somethings as they chop vegetables destined for a giant pot of soup already steaming on the stovetop. Two well-worn sofas line the windowed front wall of the room. A couple is entertaining their young son with a board game. At the opposite end of the room are overflowing bookshelves and some easy chairs. An older woman ensconced in one appears to be sleeping, although later, when I tell her I envy her ability to nap in the midst of such commotion, she will chide me that she was meditating. In the middle of the room is a wooden table lined with an assorted collection of stools and chairs, including some Windsor chairs handmade by Larch.

The soup is ready, and it's time for dinner. More guests arrive, a total of sixteen. The dinner is a coconut, fish, and vegetable soup flavored with *Saccharina,* as well as rice and various salads. All are delicious. I listen to Larch, the apprentices, and the guests talk, their voices overlapping. They discuss progress on the new drying house, a harvesting trip two apprentices are planning for two in the morning, and the program at a Rockland yoga center. Daniel, one of the apprentices, brings to the table a black, coruscated lump the size of a man's fist. It is a piece of chaga, a fungus that grows on birch trees, and people use it to make a tea that some claim supports the immune system. I instinctively decline a proffered cup of chaga tea, but later have second thoughts. I've always liked mushrooms — the stronger tasting, the better. Is a fungus tea any different from a sautéed chanterelle? Most Americans turn up their noses at edible seaweed but think nothing of eating spinach. I'm reminded that culture dictates which foods we find palatable.

I have a choice about where to sleep, so I choose a bed in the office across the hall rather than one in a tent. (I'll take easy access to a bathroom rather than a stumble through the woods in the middle of the night.) Early the next morning, I join the four apprentices in one of the drying houses. They're all wearing jeans and old T-shirts streaked and stained ochre from the high levels of iodine in these and all brown seaweeds. They hoist and drape kelp blades over wooden rods, like hanging towels over a clothesline. The July sun shining through the transparent roof banishes the chill in the air. With the sun's heat and air flowing from large fans, the kelp will be dehydrated in forty-eight hours. If the weather turns cloudy and cool, as it can easily do in summer here, the drying house will have to be heated. Outside each house is an old decommissioned oil tank that, connected by a pipe and fired by logs, serves as a furnace. Another part of an apprentice's job: get up in the middle of the night and feed the furnaces.

Nina joins me in the drying house and hands me a gift of a small package of dried, crumbled seaweed soup mix. If I protect it from light and humidity, it should keep for at least two years. (The package includes a recipe, and there are more on the company's website.) I say my good-byes and, feeling a bit guilty, drive off toward Portland to visit one of Larch's aquaculture competitors. I admire Larch's commitment to responsibly harvesting wild brown seaweeds, but the truth is that they hold far too much promise as a rich, sustainable food source for us to rely on the limited natural supply. I want to see how this might work on a larger scale.

6

Flash!

I've found the street — Presumpscot — but drive right past the office of Ocean Approved. The business is easy to miss; it operates out of an ordinary white clapboard house and there is no sign out front. On my second pass, I pull into the driveway to park on the grass next to two white cargo containers in the side yard. I have barely stepped out of my car before Tollef Olson, who founded the company in 2006, is at my side, talking about kelp. I plead with him to wait while I rummage in my backpack for my voice recorder. As soon as I have it in hand, Tollef shepherds me toward an old brown station wagon.

"Just push that stuff onto the floor," he says of the tangled line and small anchor on the seat, as he proceeds to talk fluently, emphatically, and nonstop. We're headed, he tells me, to Southern Maine Community College, where he has his nursery and lab. Observing him as he drives, I see a middle-aged man, his brown hair graying at the temples, burnished and lean in face and frame. Larch was preternaturally mellow; Tollef is equally, but oppositely, incandescent. It's a fine September day, so the windows are down, and I only hope my recorder will capture his voice over the noise of the engine and the rush of air through the open windows.

Tollef (it's a family name, he says, passed down for generations from his Norwegian ancestors) grew up in Auburn, Maine, in the 1950s and '60s. His family lived by a lake, where he learned to water-ski and ice-skate, but his love has always been the ocean. In 1970, the Olsons moved to the west coast of Florida for his mother's health, and when he wasn't in school, he was working part-time in local restaurants or on commercial fishing boats. At seventeen, he dropped out of high school and took to the sea.

"Marine salvage — that's the fancy word for treasure hunting — and commercial fishing was mostly what I did, as well as surfing and just enjoying life. I shrimped off the Florida Keys and gill-netted and long-lined from Maine all the way down to British Honduras."

In 1980, he moved back to Maine, to Bar Harbor, where he and his brother opened a restaurant that specialized in Asian-American fusion cuisine. In winter, when the restaurant business was slow, he'd sign on for salvage trips, diving to wrecks off the coasts of Australia and South America, and ultimately crossing the Atlantic, Pacific, and Indian Oceans on boats.

"The whole time we had the restaurant, I'd been interested in aquaculture, both as a buyer and because I just thought it made sense. I'd been involved in a lot of fisheries — swordfish, mackerel, mullet, shrimp, to name a few — that had gone boom and bust. I'd seen what happens: people gear up, fish out all the product, and the next thing you know, there's no industry. Fish farming, done right, I could see makes sense."

Nonetheless, after selling the restaurant in 1986, Tollef jumped into the budding North American sea urchin fishery. In Japan, people traditionally eat the spiny creatures' uncooked gonads, a sushi delicacy called *uni*. At the time, the yellow-orange, tongue-shaped organs were selling for more than $100 per pound at restaurants. Sea urchins had always been a scourge to Mainers — they eat the bait out of lobster traps, clog the traps, and mow down the kelp that protect immature lobsters and fish from predators. In the 1980s, however, Japanese importers recognized that Maine's green sea urchins were not only nearly identical to the East Asian species but, thanks to a favorable exchange rate, cheap. Maine urchins suddenly had a market value. And not only that, lobstermen could harvest them in winter, just when they needed another source of income.

The pest became a prize. Commercial landings rose from nearly zero in 1986 to 41 million pounds in 1993. The urchins were so ubiquitous in Maine waters, no one imagined they could ever be overharvested. But they were, and when the population crashed, so did the fishermen's profits. The catch fell to 11 million pounds in 2001 and was only 1.5 million pounds in 2016.

Tollef, however, had long since moved on. In 1997, he started his first aquaculture business, Aqua Farm, raising mussels. Wild mussels grow in intertidal zones, where they attach themselves with their thready "beards" to rocks and to each other, forming clusters. Traditionally, harvesters have collected mussels from the shore with a rake or from a boat with a drag, but Tollef had another idea. He leased a site off Bangs Island from the state and moored a floating platform of beams. From the beams he hung weighted mesh ropes seeded with spat, which are pinpoint-size baby mussels. His location was ideal, protected from storms and with just the right amount of current to feed his mussels a good diet of plankton. Unlike wild mussels that are exposed to air at low tides and therefore unable to eat for hours, his mussels constantly filtered the water for food. Hanging from ropes in relatively calm water, they put more energy into growing meat and less into building thick protective shells. Consequently, they matured quickly and remained tender. As a bonus, they collected less sand and silt because of where they grew, which made them cheaper to process.

Inevitably, kelp grew on Tollef's mussel-covered ropes. He had long been a fan of cooking with dried kelp, both for his family and at his restaurant. "Not only does it add flavor," he says, "it is literally a multivitamin with fiber. That's why Ocean Approved has trademarked the phrase 'Kelp, the virtuous vegetable.' And it's virtuous not only because it's arguably the planet's healthiest vegetable, but because it takes no fertilizer, insecticides, or herbicides. And you don't need any arable land or fresh water to grow it." So, whenever he was out on his mussel platforms, he took seaweed home with him. But instead of drying it, he began experimenting with cooking it fresh.

"Kelp is just like peas. Peas are very good and healthy when you dry them, and they make great soup. But if you get a fresh pea and compare it to a dried pea, it's not the same at all. A fresh or fresh-frozen pea is sweet, it's green and crunchy." He shrugs. "Don't get me wrong, I like dried seaweed, but I'm a cook, and you have a lot more versatility cooking with fresh or fresh-frozen seaweed."

It was relatively easy to get people to try adding a little crumbled dried seaweed to a dish; dried seaweed looks much like dried herbs. Convincing people to eat seaweed that actually looks like seaweed was another matter. Tollef, however, thought the time was right. The foodies had arrived, with their interest in nontraditional dishes and offbeat flavor combinations. At the same time, the organic food phenomenon was taking off, and seaweeds from Maine's clean waters were free of contaminants. Baby boomers were confronting their mortality and increasingly interested in eating healthfully. Plus, the number of sushi restaurants in the US had quintupled between 1988 and 1998. Americans, he thought, having grown accustomed to eating wakame in miso soup, hijiki and wakame in seaweed salads, and nori in sushi, might be ready to expand their culinary horizons.

Tollef knew that to make a success of a fresh seaweed business, he couldn't simply sell it seasonally to local restaurants—that would seriously limit the scope of his enterprise. But flash-freezing, a technology that he understood well from his commercial fishing days, was an intriguing possibility. When you freeze a fish in a kitchen freezer, the water inside its cells freezes gradually and forms ice crystals. The crystals are sharp enough to pierce and collapse the cell walls, and the result is a mushy cod or salmon when thawed. Flash-freezing, on the other hand, reduces the temperature of fish to −40 degrees Fahrenheit (also −40 degrees Celsius) in a matter of minutes. When food is flash-frozen, the water transforms into a solid so quickly that crystals don't form, which means the cells are undamaged and vitamins are less likely to be degraded. Fish flash-frozen at sea will often be tastier than a fresh fish that takes a couple of days to get to market. Tollef figured it was worth trying to flash-freeze seaweed and, with the help of grants from the Maine Technology Institute, he mastered the technique.

With that problem solved, there was the question of sourcing the seaweed. For the scale of operation Tollef had in mind, harvesting it wild would not be sufficient. He figured he would need to farm his kelp.

We arrive at the Southern Maine Community College and head up some concrete steps, into the warehouse-like building that is home to Ocean Approved's labs and seaweed nursery. I was expecting large tanks, but Tollef shows me four ordinary tabletop aquariums. Three are empty but the fourth holds several of what look like two-foot-tall spools of string standing on end. In fact, that is exactly what they *are*, each one with four hundred feet of string wound about it. These spools are for research, but later in the season the aquariums will hold twenty-eight spools at a time, meaning that, altogether, there will be over a mile of string in his fish tanks.

A bank of grow lights shines on the aquarium from one side and natural light streams in from the other. The strings are light brown and look slightly fuzzy. Tollef is pleased. A few short bits stick out from the spools and he reaches in and clips a piece, then puts it on a glass slide and drops a cover slip on top. I follow him to a nearby office, where he puts the slide under a microscope. "Oh, these are beautiful," he crows. "They would definitely be ready for open water. They look like plants now. You can even see the darker area down in the base of the stipe. That's where all the growth comes from."

I look through the microscope. Indeed, there is a Lilliputian sugar kelp with its stipe, and single blade. The holdfast is still too small to see but it's there, gripping the string. It's adorable.

This little seaweed and all its siblings are a month old. Tollef set up the spools in the fish tanks and inoculated the water with spores released from a piece of wild kelp he had collected. The first two weeks, he says, are always a little nerve-racking because nothing can be seen while the first phases of seaweed embryology unfold. Nonetheless, every day he turns the spools and adds a nutrient mixture. The lights go off at night to mimic the natural photoperiod. Once a week he changes the water, which he maintains at 51 degrees Fahrenheit (10.5 degrees Celsius) with a chiller.

In November, it will be time for outplanting at the company's site in Casco Bay, leased from the state of Maine. Outplanting, Tollef says, is a simple process. He starts with two surface buoys anchored to the seafloor. Below each buoy is a seven-foot length of PVC pipe, which is weighted at one end to keep it vertical. On his boat, he winds the string, now well-furred in tiny seaweeds, in a spiral fashion around a one-inch-diameter line, which is then stretched between the lower ends of the PVC pipes — that is, seven feet below the surface. The depth allows enough sunlight to reach the kelp while ensuring that even large sailboats with deep keels can safely cross above. The baby kelp grow at a prodigious rate, as much as four inches per day during the peak growing season. When, four to five months later in early spring, a mechanical hoist pulls out the line, it will be heavy with densely packed, five-foot-long kelp.

Another reason for farming kelp instead of wild-harvesting it is control. All his seaweeds are growing under identical conditions and are easy to monitor. "I need a steady supply of high-quality product, and the best way to get that is to grow it myself. Besides, right now Maine has a lot of wild seaweed, but we've already seen a little bit of overlap between harvesters. If you have multiple divers on the same kelp, you start degrading the biomass to the point where it's not sustainable anymore. People like Larch are very conscientious, but not everyone has his integrity. It's just human nature. If you're not getting what you hoped from your beds — maybe there's been storm damage — it's real easy to cut too much to make sure you can pay your bills that season. I would hope there's always going to be some wild-harvesting, but if it's not managed properly, which seems to be the history of most fisheries in the US and around the world, we're going to see problems."*

• • •

*Ocean Approved has received funding from NOAA and the Maine Technology Institute to develop its technologies. (Larch was correct that seaweed aquaculturists have gotten government financial assistance.) The rationale is that kelp aquaculture promises to provide jobs in a new industry conspicuously gentle on the environment. Sounds good to me.

We drive back to where I parked my car, which is where the company's office and processing operation are. Tollef shows me how the brown-gold blades of *Saccharina latissima* are blanched, a process that turns them bright green, and then cut into linguini-like strips. The stipes are sliced into hollow rounds. He opens two frozen packages and thaws the seaweed under warm water. I taste some *Alaria*, cut into shreds, which Tollef says makes good salad. *Digitata laminaria*, crunchy and mild, is sliced thinly for a seaweed coleslaw. The company also makes puréed cubes, ready to pop into a blender for a smoothie. All are flash-frozen and packaged, and then shipped in a Styrofoam container.

The fresh product opens possibilities for creative new dishes, and Tollef has worked with chefs at Johnson & Wales University to create new recipes, all of which can be found on the Ocean Approved website. He sends me home with samples of Ocean Approved products. (I can report that ocean perch sautéed with shreds of sugar kelp in butter, lemon, parsley, and garlic is delicious.) Ocean Approved has a steady demand for its products from restaurants and institutions, and it has plans to sell directly to consumers. The company has published a manual, available free online, on how to cultivate kelp, and it has expanded its capacity by buying from a dozen independent kelp farmers along the Maine coast.

Tollef continues to push the boundaries of seaweed cuisine. He recently moved on from Ocean Approved and, partnering with Treetops Capital, started a new seaweed company called Ocean's Balance. While still a big fan of flash-frozen seaweed, he sees an additional market for a shelf-stable puréed product. A kelp purée can be easily incorporated into recipes for sauces, burgers, soups, and other dishes where it ups the nutritional content. The company markets both to consumers — a grocery store not far from my house carries small jars of the purée — and to institutional food services.

Pushing myself to my culinary limits, I recently made tomato soup according to a recipe on the Ocean's Balance website, both with and without two tablespoons of kelp purée, and subjected six dinner guests to a blind taste test. My tiny sample size hardly makes this a definitive study, but five of six preferred the soup with kelp. Recently, no less an authority than Sam Sifton, food editor of the *New York Times*, wrote a piece for the newspaper's magazine in which he recounted eating steamed bluefish at Houseman, the Manhattan restaurant of chef Ned Baldwin. Sifton had low expectations for the dish, but, he writes, "Baldwin's fish was fantastic, almost exploding with flavor: briny, buttery-rich, silky-salty, with a powerful roundness." The secret of the dish? Dulse, another seaweed native to both sides of the North Atlantic, that Baldwin blends with butter and slathers on the fish. Sifton was so taken with dulse's flavor-enhancing ability, he wrote, that he now cooks with the seaweed all the time.

Will Westerners ever eat as much seaweed as East Asians? Ten years ago, I would have laughed at the notion. But I think of those kindergarteners showing up at school with seaweed snacks – delicate, lightly salted, slightly crunchy yet melting on the tongue – in their lunchboxes. Chefs, grocers, and growers are giving seaweeds a much-needed nomenclature makeover, calling them "sea vegetables" or, even better, the cuddly "sea veggies." (The branding is a work in progress. I've seen them called "the kale of the sea," which I still find less than enticing. Whole Foods calls them "the pearls of the vegetable family," which is odd since pearls aren't edible. More creatively, Pembrokeshire Beach Food calls its new pepper dulse product "sea truffle.") I have no doubt that there will be more savory dulse and kombu and wakame, crispy nori, and other sea treats coming to dinner tables in the West. Use them powdered as a spice. Eat them raw in salads. Add them dried, puréed, and flash-frozen to soups, stews, quiches and omelets, and baked goods. And watch for them slipped into processed foods to raise the nutritional profile. By any name, seaweeds are headed to a meal near you.

7

Spirulina

Seaweeds are not the only algae people eat. In China, people have long gathered the dark green, hairlike cyanobacterium *Nostoc flagelliforme* from the ground in the high deserts of China and Mongolia. Desiccated and dormant for much of the day, *Nostoc* "wakes up" when moistened by dew. It's been a food source in the region for two thousand years; the Vietnamese have acquired a taste for it more recently. Known in Chinese as fat choy, it's served most often on holidays, as a symbol of good luck. Fat choy doesn't have much taste, but it absorbs flavors from other ingredients, much the way cellophane noodles made from vegetable starch do.

Another cyanobacterium, spirulina, has been on the menu in Latin America and Africa for hundreds of years, and possibly much longer. Under a microscope, spirulina looks like skinny green corkscrews, hence its name. In the 1500s, Spanish conquistadors arrived in the Aztec capital city of Tenochtitlán, which was built on an island in an enormous lake. (The lake, long drained, is now the site of Mexico City.) The invaders noted how the Aztecs used fine nets to gather a blue-green substance from the lake's surface, and they reported that vendors were selling "small cakes made from a sort of ooze which they get out of the great lake, which curdles, and from which they make a bread." In Central Africa, people have long harvested spirulina from the shallow pools of water that form when Lake Chad overflows during the rainy season. Women of Chad's Kanembu tribe collect spirulina from Lake Kossorom in clay pots, strain it through cloth, and dry it in the sun. Cakes of *dihé* are crumbled and mixed into about 70 percent of Kanembu meals.

• • •

Today in the United States, spirulina, grown outdoors in artificial ponds and processed into a deep blue-green powder, is a business. Earthrise Nutritionals, Inc., one of the world's largest producers, grows the cyanobacteria in forty-seven acres of raceway ponds in the middle of California's Sonoran Desert, a few hours southeast of Palm Springs and thirty-two miles north of the Mexican border.

When I pull up to Earthrise at 10:30 on a mid-July morning, the outdoor temperature is already 110 degrees Fahrenheit (43 degrees Celsius), which makes it a perfect day for spirulina, if not for humans. Dr. Amha Belay, senior vice president and chief technology officer, meets me at the door and suggests we take a quick tour immediately, before the temperature rises even higher. Dr. Belay is a soft-spoken man with a café au lait complexion and the lilting accent of his native Ethiopia. He came to the company on a sabbatical from Addis Ababa University about thirty years ago and has been at Earthrise ever since, nearly since the company's founding.

We climb an outdoor steel staircase about six stories tall, and I learn the hard way why it's not a good idea to touch, much less grab, an iron railing in the desert. The staircase hugs the side of a sixty-foot, rocket-shaped tower that houses one of the plant's two spray dryers. There's a panoramic view of the facility from the top. Below, to the left, are two artificial ponds filled with water drawn from the mineral-rich Colorado River. The water rests for a few days in the smaller pond to allow sediment to settle out, then it flows by gravity to the larger pond, where it's supplemented with soda ash (sodium carbonate). The soda ash raises the pH level of the water considerably, making it over a thousand times more alkaline than tap water, which suits the spirulina just fine while discouraging many other organisms looking to share a home with or make a meal of them.

The water is then pumped as needed to thirty-seven "raceway" ponds. Raceways are artificial, shallow bodies of water that look like long skinny racetracks with an exceedingly narrow infield—just a partition, actually. A paddlewheel that spans one side of the track moves water continuously around the circuit. From up here, Earthrise's ponds look like black lozenges neatly arranged, side by side, startling in the empty ochre landscape.

Spirulina are prolific. During the April-to-October growing season, they divide rapidly, increasing in mass by 30 percent each day. Every two or three days, when their concentration hits a certain mark, workers — onsite around the clock — pump about 50 percent of the pond's water through underground pipes to the processing building to our right, a building to which we now happily retreat. Here the pond water passes through a series of stainless-steel screens that concentrate the biomass into a slurry and then into a paste. (The residual water is piped back to the pond.) The spirulina paste, which I see oozing out of the last machine in the line, falls in languid folds like emerald-green cake batter onto a conveyor belt. The belt transports the batter to the drying tower, where, sprayed as fine droplets at a very high temperature, 93 percent of its water evaporates. The residual fine dark green powder is sterile and can be safely stored for more than a year.

Half of the five hundred tons of spirulina that Earthrise produces each year are consumed as a dietary supplement. Most of the remainder is sold to food companies like Starbucks that include a bit of spirulina in their bottled fruit smoothies. Recently, the FDA approved the use of phycocyanin, the blue pigment that deepens spirulina's hue to emerald, as the first natural blue food coloring. The approval opened up a new market, and now Earthrise and other growers extract and sell phycocyanin to candy and other food manufacturers that want to replace artificial blue pigments.

Spirulina growers often advertise their product as a "superfood" and proclaim an astounding nutritional profile: spirulina, they say, has 2,300 percent more iron than spinach, 3,900 percent more beta carotene than carrots, 300 percent more calcium than milk, and 375 percent more protein than tofu. But don't be misled. Numbers like these are on a per-calorie or per-gram basis. On a *per-serving* basis, it's a different story. Earthrise and other companies advise consumers to eat just 3 grams, or about a teaspoon, of spirulina per day. A serving of spirulina does contain 140 percent of the daily recommended value of beta carotene, a pigment that our bodies convert to vitamin A. One teaspoon also provides about 10 percent of our recommended daily amount of iron, and it has antioxidant qualities. But a serving has less than 2 percent of the recommended daily allowance of protein.* If you ate five times the recommended amount – which would mean swallowing 30 half-inch spirulina tablets – you would get as much protein as you'd find in two eggs.

Even if you did eat that much spirulina, your body would limit the amount of beta carotene it converts to vitamin A (too much vitamin A damages the liver). While eating excessive amounts of the pigment is therefore not toxic, consider that excess beta carotene accumulates in the outer layer of our skin, which means that if you eat too much of it and you have pale skin, you may turn orange. That might be interesting at Halloween, but it probably isn't a look you'd want year-round, and it can take months for the condition – carotenosis – to disappear. (You could add to the colorful effect by eating spirulina powder, which will leave you with a temporarily blue-green tongue and teeth, as well as a grimace from the taste.) Bottom line: If you feel you are lacking in vitamin A, you can certainly swallow six tablets of spirulina daily. But you could also simply eat one-third of a carrot per day and get 100 percent of the vitamin as well as a host of other micronutrients at a small fraction of the cost.

This is not to say there is no place for spirulina in the human diet. Most people in developed countries get sufficient vitamin A simply by eating spinach, carrots, and other fruits and vegetables, or vitamin-fortified foods like many breakfast cereals. But in developing countries, about a third of children under the age of five don't have access to these foods and consequently are vitamin A deficient. About 250,000 to 500,000 children will go blind from vitamin A deficiency, and half of those children will die. A daily teaspoon of spirulina could save their vision, or even their lives. Pregnant women who lack access to these foods deprive their fetuses of vitamin A, so a teaspoon of spirulina could be a critical prenatal supplement.

---

*For vegans and vegetarians inclined to take spirulina as a protein substitute, note that spirulina does not contain the form of vitamin B_{12} that humans can absorb and may actually aggravate vitamin B_{12} deficiencies.

The iron in spirulina can also help with anemia, another common health problem for many children in the developing world.

A number of international and local aid agencies distribute emergency vitamin supplements in developing countries to combat vitamin deficiencies, but a Swiss charitable organization, the Antenna Foundation, has taken more of a teach-a-woman-to-fish approach. The foundation instructs women on how to grow spirulina in tubs or shallow, artificial ponds. The villagers stir the pond water manually or with a simple motor, and they dry the cyanobacteria in the sun or in solar dryers. (One advantage of sun-dried spirulina: unprocessed, it has little or no taste.) Not only does the product supplement children's diets, but the women can package and sell it for much-needed cash. Spirulina needs less water and fertilizer than vegetables, tolerates high temperatures well, and can be grown and harvested year-round in many tropical countries.

Experts from Antenna have helped villages in India, Cambodia, Laos, Nepal, and seven African countries from Mali to Madagascar start and maintain spirulina ponds. In recent years, other charitable organizations have started similar programs. Some continue to oversell the product (one group, for example, claims that "one teaspoon has the nutritional value of several servings of common vegetables"), but in countries where people are malnourished, spirulina should have a place at the table.

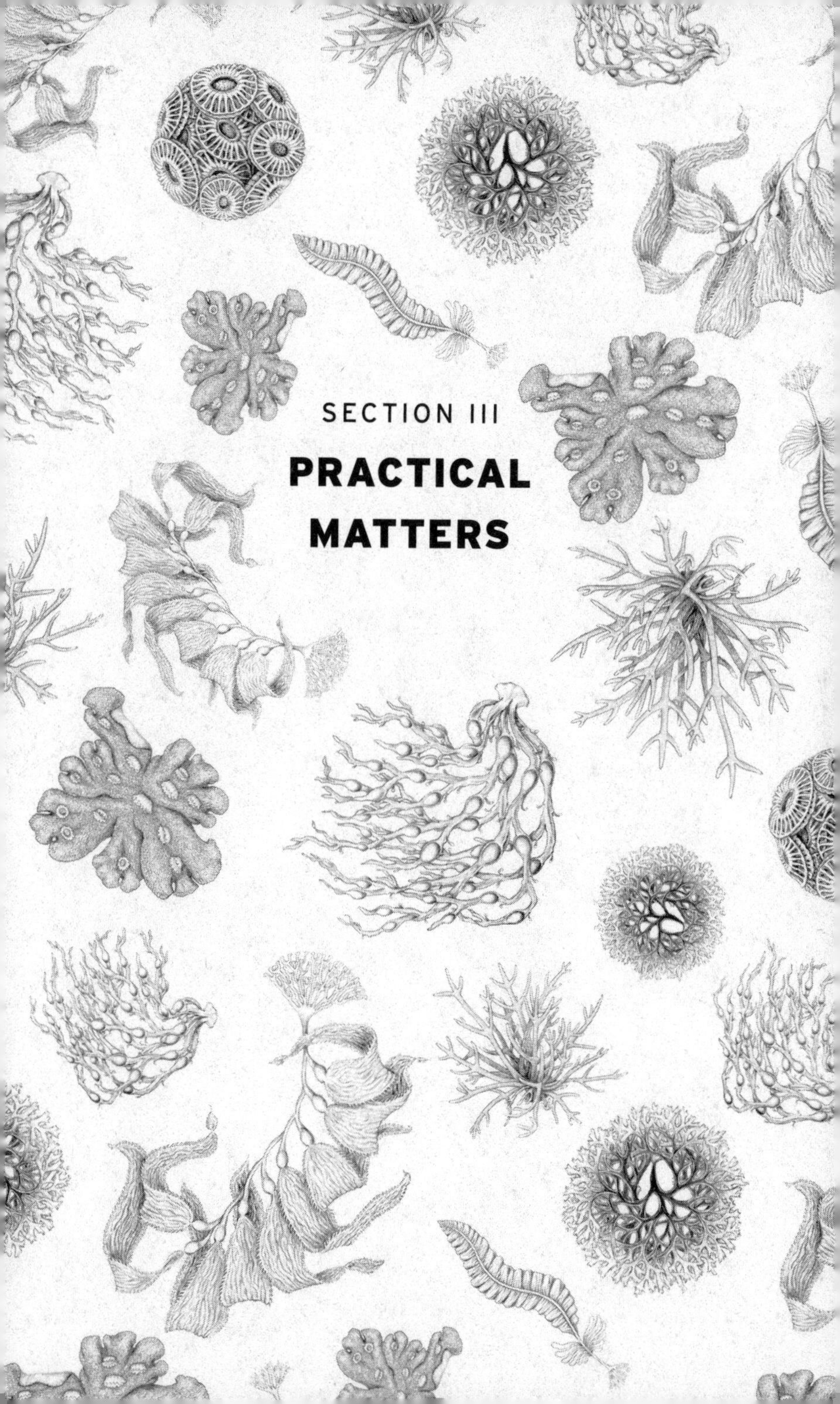

SECTION III

PRACTICAL MATTERS

1

Feeding Plants and Animals

Algae are dynamos; they crank out all kinds of nourishing organic compounds. But we needn't eat algae to benefit from them.

One of the first ways people put algae — macroalgae, specifically — to work was to build or restore the fertility of farmland. For centuries, Irish smallholder farmers living by the sea collected seaweeds at low tide from the rocky shores. They loaded their gatherings, wet and heavy, into large woven baskets and took them home strapped across a pony's back, or their own. There, they rinsed them, set them aside to decompose, and, weeks later, spread them on their fields and mixed them in the soil of their raised potato beds. Seaweed fertilizer was sometimes the difference between a farm's failure and a family's eviction or another year on the land. On the limestone Aran Islands off the west coast of Ireland, inhabitants over the generations hoisted and spread so many tons of seaweed and sand from the beach that they created farmland where none existed before.

Farmers have also been feeding seaweed to their animals for at least two thousand years. A Greek writer noted in 45 BC that "in times of scarcity, [farmers] collected seaweed from the shore and, having washed it in fresh water, gave it to their cattle and thus prolonged their lives." Irish stockmen have long mixed it into pig, cow, and sheep feed, and not just in hard times. Usually, seaweed makes up only a small percentage of the animals' feed, but on one of Scotland's Orkney Islands, a breed of wild sheep has been living primarily on seaweed since 5000 BC. The North Ronaldsay sheep, which stand eighteen inches at the shoulder and weigh about forty-five pounds, graze along the rocks at low tide — and not just out of desperation. They thrive on their diet of macroalgae and turn up their petite muzzles when grass is offered.

The seaweed in the Irish farmers' baskets was often *Ascophyllum nodosum* (also known as rockweed or bladderwrack), a brown species that looks like a mop of golden brown, stringy stipes and many small blades. The species can grow to be eight feet in length and individuals can live as long as fifteen years. It is ubiquitous on the coasts of Ireland and other Northern European countries, and it blankets miles of shoreline in Maine, New Brunswick, and Nova Scotia. Nova Scotia is where I've come to investigate what this common, weedlike macroalgae has to offer the world.

My investigation begins in a small white skiff that is puttering slowly through the morning fog across the Wedgeport Harbour on the province's western shore. At the helm is Jean-Sebastien Lauzon-Guay, a marine biologist for Acadian Seaplants Limited. Acadian is the world's largest producer of dried rockweed, which it processes into a liquid fertilizer for crops and a dry ingredient in feed for farm animals and pets. The privately held company employs about three hundred fifty people directly and contracts with about six hundred independent harvesters in Canada, the US, Ire-

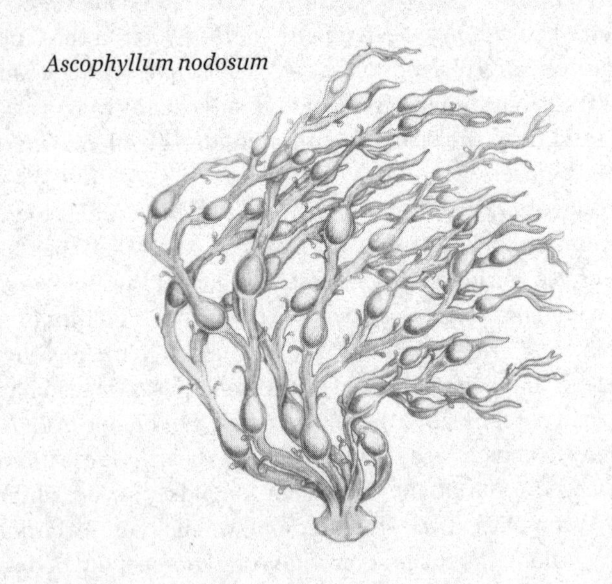

Ascophyllum nodosum

land, and Scotland to produce millions of pounds of the seaweed. Acadian ships its products from Canada to eighty countries around the globe; it recently bought and is expanding two processing facilities in Ireland and Scotland.

It's noon and about ninety minutes after low tide, and Lauzon-Guay is taking me to meet harvester Jeff Doucette, who is getting ready for his second harvesting trip of the day. We head to the far side of the small harbor, where a half-dozen seaweed skiffs are tied up to a wharf. The tides here are dramatic, rising and falling at about a foot per hour, and, at the moment, the boats are fifteen feet below the wharf. Pilings, algae-covered and sturdy and broad as tree trunks, loom behind and above the boats, dwarfing them. Lauzon-Guay points out Doucette's boat, which — forgive me, Jeff — looks like a pitch-black, cast-iron bathtub. The tub has three feet of freeboard now. In a few hours, though, when it is piled high with seaweed, the gunnels won't be far from the surface. The *Black Out* is no beauty, but she is well made for her purpose.

Doucette, standing in his boat and holding a long metal file in one hand, greets us with a generous hello as we pull up alongside. With a round bald head and a linebacker's physique, Doucette also looks purpose-built. He's been harvesting seaweed for thirty years, he tells me, long enough to see his son grow up and buy his own boat. ("He used to come wit' me," he chortles, "but Dad was a hard boss." Acadian French is Doucette's native language; he speaks English with a lively inflection and has no use for the dental fricative.) We've arrived just as he's finished sharpening his rockweed rake, a tool that no more would double for a leaf rake than a lumber saw would substitute for a steak knife. For one thing, it's twelve feet long. At the business end of the rake are three steel components, each one welded about six inches from the next. Lowermost is a pair of curved runners that keeps the blade about five inches off the bottom, so the seaweed won't be cut too close to regenerate. Next is a wide and wickedly serrated blade that cuts the seaweed cleanly. Uppermost is a row of closely spaced long and narrow tines that grab the cut seaweed so it can be lifted out of the water.

"I do my own rake," Doucette tells us. "I probably tried ten different styles, moving the blades, changing the shape. Every little t'ing helps on the load." He laughs, "Dat is, if your back and shoulder can take it."

Doucette goes out twice a day on the flood or the ebb, usually accompanied by other harvesters in their own boats. The trick is to get to the seaweed when the tide is about halfway in or out. If the water is too high, the rockweed is covered too deeply; if the tide is too low, it's out of reach on the exposed rocks. No one harvests seaweed from land anymore, at least not in Nova Scotia. That's slower and harder work and, worse, it clear-cuts the rockweed in a single area. Once an area has been clear-cut, it won't be repopulated for many years. With the seaweed rakes that Doucette uses, the buoyant seaweeds are not dislodged. In their drifting boats, the harvesters spread their efforts out across a wider area than a person on shore could.

Provincial regulations limit the total harvest to about 20 percent of the available biomass. It is Lauzon-Guay's job to track the take from each of the dozens of sectors that the various provincial governments lease to Acadian Seaplants, and to close sectors to harvesting when the maximum is taken. The company has been making a business from rockweed for decades. It's clear to me that it is committed to ensuring that the ecosystem, which provides critical habitat for young fish and seabirds, is maintained. According to Doucette, the crop is more abundant now than in the old days, when seaweed harvesting was a free-for-all.

When I ask what a good day's haul is, Doucette tells me about eight tons, raked in the course of two outings. I'm amazed at the idea of lifting that much weight, but he scoffs, "You only lift t'irty to fifty pounds at a time." For many people, seaweed harvesting is seasonal work from early summer through the fall, when there's no lobstering, but Doucette works year-round at it. Only when his rake handle is so coated in ice that it slips from his grip does he quit.

Doucette casts off and we follow the *Black Out* into the fog. We're headed toward Tucker Island, a flat uninhabited island about a mile and a half from the harbor. As we approach the shore, the boats slow to a near idle, and we're drifting slowly in shallow, dark water dotted by clumps of rockweed.

Standing midship, Doucette throws his rake its full twelve-foot length and lets the steel end sink to the bottom. He slowly draws it in, hand-over-hand. He levers up the rake head and, with a deft swoop and twist, drops what would constitute an armful of wet seaweed—in autumnal hues of brown, olive, and gold—into the bottom of the boat, and then he tosses out his rake again. He works at a steady, unhurried pace, and explains that he generally catches and cuts two or three clumps of seaweed in a single pass. Newbies tend to use a shorter, lighter rake, but, he says, for about the same amount of effort, they get only half as much "weed." I'm offered the chance to try it myself, but I can imagine that his rake, weighed down at the far end with a load of seaweed, could easily lever me overboard, so I politely decline. Instead, Lauzon-Guay suggests we motor to the shore to get a closer look at the rockweed.

The shore is completely blanketed with mounds of the stuff; it's solid seaweed from the water's edge to the forest's edge a hundred yards away. At first, I think the seaweed itself forms mounds, but as soon as I step out of the boat, I realize that jagged rocks underlie them. It's hazardous walking. (Lauzon-Guay tells me—a little solace in advance?—that whenever he brings interns to the shore for the first time, they always fall in.) We squat down to take a closer look. Each plant has twenty to thirty strands, and the strands sport short branches and olive-size, olive-drab bladders. The bladders, one for each year of growth, help keep the fronds afloat, like bobbers on fishing lines. Parting one of the mops, Lauzon-Guay shows me where it's attached to the rock. Although the holdfast glue is powerful, winter storms and thick ice will eventually rip the larger seaweeds off their foundations.

I manage to get back into the boat with cold water in only one of my knee-high boots, and we head back to the harbor. I turn and shout a thank-you to Jeff, who has drifted a short distance away, but he is dissolving in the fog and I don't know if he hears. At the dock, Lauzon-Guay points out trailer-size blue boxes on the wharf. Acadian provides a box for each harvester, and on his return Doucette will use a hoist to transfer his booty into the box. When the box is full, an Acadian truck takes it away, weighs its contents, and replaces it with an empty one. His seaweed will travel either to the company's Cornwallis or Yarmouth facility for processing.

Farmers have known for centuries that incorporating seaweed into soil makes for more fruitful crops, and, until recently, they assumed the algae was acting as a fertilizer, providing mineral nutrients. That's true, but research now reveals that the story is more interesting and complex. Seaweeds contain certain compounds, known as biostimulants, that turbocharge plants by other means.

I know little about biostimulants or the role they play in plant growth, so I meet with Jeff Hafting, senior manager of cultivation science, for a tour of the company's research and development facility in Cornwallis. Hafting—tall, blond, and lanky—earned a PhD in botany from the University of British Columbia and then joined a startup in Hawaii that cultivates abalone, the sweet-tasting, marine snails with fist-size shells that are especially valued in East Asia. The abalone were fed with dulse, and he developed a method for growing, in outdoor tanks, the millions of pounds of seaweed necessary to feed the hungry gastropods.

Here in Nova Scotia, Hafting's responsibilities range from growing seaweed on land (which we'll get to later) to assessing the biological activity of rockweed. For that, we start in a lab where tiny mung bean seedlings are growing in clear glass bottles no more than two inches tall on shelves intensely lit by grow lights. Hafting plucks out three of the little bottles and focuses my attention on the plants' laterals, the thready roots that grow sideways off the sturdier vertical radicle. The first bean plant is growing in clear distilled water, and its laterals look like a two-day, white stubble on the radicle. The second seedling is growing in distilled water supplemented with just the mineral nutrients extracted from the rockweed. This seedling's lateral roots extend like branches of a Christmas tree.

The third seedling is growing in rusty-looking water that contains both the mineral and the bioactive components of the seaweed extract. At first glance, its root system looks the same as the second seedling's, but when I peer more closely I see that the laterals cover a greater percentage of the radicle and grow more densely. It's not a knock-your-socks-off difference, but these seedlings are only about a week old. Besides, Hafting explains, the most important difference can only be seen under a microscope. Only with magnification would I be able to see the root hairs that encircle, like a bottle brush, the tips of all the lateral roots. There are more root hairs on the bean plant growing in the full seaweed extract than on the others. (The root hairs — along with their symbiotic mycorrhizae — are responsible for taking in most of a plant's nutrients.) When a plant treated with biostimulants matures, its entire root system will be more luxuriant, and it will flower and fruit more prolifically.

We walk through a series of small greenhouses and rooms illuminated by grow lights, and stop in one where young plants in black pots — green peppers with walnut-size fruits and tomatoes in bloom stage — are lined up rim-to-rim. Some of the plants are fed with rockweed extract and others are not, and their growth is measured. In another room filled with three-foot-tall corn plants, Hafting tells me just a little sheepishly, "This is a sort of torture lab for plants. We may add some salt to the water we're watering with. We may do some drought trials. With the extract, we find that plants overcome these stresses better, mainly because of the strong root systems they have. We've even done things on cut flowers. They last longer because their antioxidant content is higher."

The extract works in several ways. For one, it encourages plants to exude polysaccharides that feed mycorrhizae. The better fed the fungi are, the more numerous they are; the more numerous they are, the better nourished the plants are. Bonus point: the gel-like exudates give soil a texture that holds water better. The extract also stimulates the expression of plant genes that control carbon fixation, mineral metabolism, detoxifying enzymes, osmolytes (molecules that control water uptake), and growth hormones.

Lab and field tests prove that applying small quantities of seaweed extract to soil or plant foliage can improve crop yield by 10 to 30 percent, although the cost of the extract makes it practical only for high-value crops. Sixty percent of California table grapes, for example, are now grown in soil treated with seaweed extract. The seaweed-as-biostimulant market is currently $450 million, but the industry is still in its infancy. Given that crop stressors, including drought, increased water salinity, and extreme temperature swings, are already a growing problem in agriculture worldwide, we'll see no shortage of demand in the coming decades. The question will be whether supply can keep pace.

Acadian's animal feed operation is seventy-five miles down the road outside Yarmouth at a decommissioned World War II airbase. When Hafting and I arrive, several large farm tractors pulling modified manure spreaders are trundling across two of the five runways, spreading loads of wet rockweed. Over the course of a day (or a few days, if the weather is uncooperative), tractors towing broad rakes called tedders will turn and fluff the seaweed several times, much as they do hay on a farm. Finally, modified silage choppers pass through to chop the seaweed into pieces. This afternoon, under clear skies, loaders are scooping up dry rockweed and depositing it in blue trucks that then rumble off to one of several old Quonset huts that remain on the base. We follow one of the trucks to the hut entrance, where a flagman wearing a fluorescent orange vest waves us inside.

Hut is hardly the word I'd use for one of these cavernous structures, except as the term implies darkness. The doors at the end of the long building are broad and thirty feet high to accommodate trucks. There are no windows, and the light fades quickly as we walk deeper inside. We watch a dump truck back up, beeping, against a hill of rockweed at one end of the building, and pour out its contents. Seaweed dust rises in the air. After the truck exits, smaller loaders shuttle seaweed to the other end of the building, where a concatenation of machinery, covered in greenish dust and roaringly loud, further dries, chops, and grinds the crop into a powder. In another operation, the pulverized seaweed is funneled into white plastic bags that each hold a ton of product. Stacked on pallets, the bags will be shipped to feed manufacturers who will add small quantities — usually less than 2 percent per volume of feed — to their products.

Until recently, Icelanders used seaweed as animal feed during the winter months when plant fodder ran out. Even when it wasn't a matter of do-or-die, coastal farmers often fed a little seaweed to their stock, maintaining that the bit of seaweed made for healthier animals. In the early twentieth century, when scientists first analyzed the content of brown seaweeds, they found mineral nutrients, fatty acids, and proteins as expected, but not in quantities that would explain any health benefit in the small amounts that farmers added to feed. They also found lots of tough carbohydrates that animals can't digest. In sum, there appeared to be no scientific reason to feed animals seaweeds, and the additive was deemed a harmless — but ineffective — folk tradition.

It wasn't until the twenty-first century that the secret of seaweeds emerged. The groundwork was laid in the 1990s, when scientists began to focus greater attention on the role of the colon (or large intestine) in animal nutrition. Most of foods' nutrients — sugars, short-chain polysaccharides, fats, and minerals — are digested by hydrochloric acid and enzymes in the stomach and small intestine. Traditionally, the colon was assumed to have two purposes. One was to reabsorb water from the slurry of food that makes it to the far end of the gastrointestinal tract. The second was to compact the remains — including the indigestible long-chain polysaccharides — and eject them from the body.

We've long known that the colon is a mecca for microbes. Unlike the stomach and small intestine, which are acidic and generally inhospitable to microscopic life, the colon has a relatively neutral pH, making it a happy home for microbes. As late as the 1970s, scientists believed that the lower gut was primarily inhabited by *E. coli*. But with better lab techniques to culture gut microbiota and gene sequencing to identify them, microbiologists have come to a better understanding of the denizens of the large intestine, both in their astonishing number (one hundred trillion, or ten times the number of our bodies' own cells) and the variety of species (more than a thousand different ones). Moreover, they realized that most of these residents are not harmful and some are highly beneficial. In fact, having a diverse community of microbes and fungi in our colon is essential to human health.

We should count ourselves lucky to be occupied by bacteria: They have digestive enzymes we lack. Their enzymes are able to break apart the toughest polysaccharides that make it to the lower gut, chopping them into short-chain fatty acids. These SCFAs can be burned by the body for energy and, we now know, provide up to 10 percent of a human's calories and considerably more of a herbivore's energy needs. Beneficial gut microbes also create vitamins and hormonelike compounds that regulate other organs. Another reason to love or at least appreciate these bacteria: They generate antioxidants and have an anti-inflammatory effect, and they make the colon more alkaline, which discourages harmful microbes from taking up residency. When the levels of beneficial microorganisms in our gut are high, there's less opportunity for malevolent bacteria to establish themselves.

The microbiota in our guts are so influential, some scientists consider them an organ. Their importance has given rise to a multibillion-dollar industry of "probiotics," concoctions of live microbes that are added to yogurts and other foods for humans and to feeds for animals. Swallowing probiotics sounds like a great idea, but the reality is that stomach acids and the liver's bile are powerfully efficient at tearing apart organic compounds, including beneficial microbes. Even if the beneficials make it through the Scylla and Charybdis of the stomach and small intestine, they are likely to arrive in the colon in a weakened state. Once there, they have to compete with the 100 trillion other bacteria for a place in the gut. For most beneficials, the journey down the throat does not end well.*

---

*Some asides on human health: There is little data to support the use of probiotics in normally nourished humans. Plus, many commercial "live cultures" are actually dead before they start the trek down the digestive tract. There is some evidence that if a general antibiotic has decimated your gut of its normal flora or if you have certain digestive diseases, probiotics can be helpful. The fad for colonic cleansing, however, is entirely wrong-headed and it wreaks havoc on the natural,

So, the question becomes, if we can't readily add beneficials to the gut, can we multiply the number of those already in residence?

That has been the objective of *prebiotics,* a term coined in 1995 by professors Glenn Gibson and Marcel Roberfroid. Prebiotics are simply the compounds that certain beneficial gut microbes prefer to eat. When these microbes have an ample supply of their favorite long-chain polysaccharides, they become more active and reproduce, to the host animal's benefit. In other words, prebiotics can tilt the playing field of the gut to favor beneficials. For years, food and feed manufacturers have added inulin, a polysaccharide derived from the chicory plant, to their products because it increases the numbers of friendly gut bacteria. In a similar fashion, rockweed's polysaccharides are favored by certain beneficial microbes in herbivores' lower guts, making seaweed a useful prebiotic.

Seaweed prebiotics may also be useful in countering the worldwide epidemic of antibiotic-resistant microbes. The problem arose because for decades beef cattle, pig, and broiler-chicken farmers have supplemented their livestock feed with low doses of antibiotics. They do so not in order to cure a disease, but to promote the animals' growth. There are a number of theories about exactly how farmyard regimes of subtherapeutic antibiotics work, but the fact is, they do. Antibiotics have been especially important in industrial farming, where animals are crowded in pens, which not only stresses them and thereby increases their vulnerability to disease, but also makes the exchange of microbes more likely.

While dosing animals with subtherapeutic antibiotics encourages faster growth, we now know that these regimens endanger human lives. Through constant exposure, animal pathogens become impervious to the medications, and when we eat meat from animals harboring antibiotic-resistant microbes, we sometimes consume those pathogens. If those pathogens proliferate inside us and cause disease, antibiotics are likewise ineffective in our bod-

---

balanced microbial ecosystem in your colon, killing beneficial bacteria and creating space for colonization by harmful ones.

ies. Even vegetarians are affected: Resistant pathogens show up on vegetables grown in proximity to farm animals. Worse, farm-bred, drug-resistant pathogens can pass along genes (through lateral gene transfer) that confer antibiotic resistance to microbes that otherwise benignly inhabit our bodies.

We are running out of effective antibiotics: Around the world, 700,000 people die every year from drug-resistant infections, including at least 23,000 in the US. The 2016 report of the Review on Antimicrobial Resistance, issued by a commission established by the British prime minister, concluded that if no new antibiotics are developed by 2050, untreatable bacterial infections could kill 10 million people per year worldwide. Few new antibiotics are in the pipeline, and of these, the same report notes, "only perhaps three have the potential to offer activity against the vast majority (≥90 percent) of the most resistant bacteria that doctors already have to treat today." The European Union banned antibiotics as growth promoters in animals in 2006. The American Medical Association is on record urging farmers to stop dosing their animals for non-therapeutic reasons, and the WHO and other prominent health organizations have called for the same action. But while major chicken producers like Tyson and Perdue have committed to reducing their antibiotic use, in the US more than 75 percent of all antibiotics administered are given to animals to encourage growth.

Although the need for the ban is clear, there is also no question the antibiotics will be missed in the meat industry. Acadian makes the case that rockweed can help replace them. (Of course, reengineering industrial farm practices to avoid inhumane crowding would be an enormous help, too.) Dr. Franklin Evans, one of the company's experts on animal physiology and one of thirteen PhDs and thirty other researchers on staff, tells me, "We've tried our product in every kind of animal, from nematodes to rats, poultry, beef cattle, pigs, rabbits, and ducks. In every case it increases the presence of beneficials and decreases *Salmonella* and *E. coli* in the gastrointestinal tract of test animals." *The Journal of Applied Phycology* published a study by Evans and a colleague that shows that kelp and other seaweeds are at least five times more powerful than inulin in boosting livestock performance and rival the efficacy of therapeutic antibiotics.

Those hardy Irish and Scottish farmers were on to something all along.

2

In the Thick of It

Today, Acadian Seaplants is heavily invested in rockweed, a brown macroalgae, but when the company started in 1967, its business was based on the red seaweed *Chondrus crispus*, otherwise known as Irish moss. The value of Irish moss, a small maroon seaweed with furcating flat fronds, lies in a polysaccharide it makes called carrageenan. Carrageenan is a phycocolloid, a term derived from the Greek, meaning "algae glue." Phycocolloids are made of microscopic, water-absorbing algal particles just the right size to stay suspended in liquid, and they have a remarkable ability to thicken, emulsify, or gel liquids.

The Irish were the first to discover carrageenan's power, perhaps while making a pot of tea from Irish moss. Carrageenan is simple to extract: just boil a little Irish moss in a lot of water, quickly add cold water, then strain and cool the jelly-like remainder. In the Middle Ages, European physicians and healers recommended it as a treatment for respiratory complaints, among many other ailments. (In this case, folk tradition was wrong; carrageenan has no medicinal properties.) By the seventeenth century, it was chiefly used in the kitchen to turn milk, sugar, and almonds into a pale, firm dessert called blancmange. When Irish and Scottish immigrants came to the US in the mid-nineteenth century, they were delighted to discover the familiar seaweed thriving on the shores of New England, and the dessert became an American staple. Fannie Farmer included recipes for a traditional version and a chocolate variation in her famous *Boston Cooking-School Cook Book* of 1918. (You can find them in the appendix.)

The dessert inspired an industry. In 1940, the Chicago dairy company Krim-ko established a factory in Scituate, Massachusetts, to extract carrageenan from the Irish moss that grew rampantly in coastal waters. The company used it to create a milk-based, pudding-like dessert called Junket, as well as to keep cocoa suspended in chocolate milk. When World War II cut off imports of traditional vegetable-based thickeners like gum arabic and xanthan gum, American companies tried carrageenan as a replacement. For thickening dairy products, nothing was (or is) better, and the compound works at extremely low concentrations, less than 0.2 percent of the volume of milk. After the war, one of the companies, the Algin Corporation of Rockland, Maine, set its chemical engineers to developing more efficient methods of extracting carrageenan from Irish moss and figuring out new ways to use it.

Algin's researchers, known as the pudding boys, cooked up a range of refined products, exploiting the three subtypes of carrageenan, as well as a similar phycocolloid made from brown seaweeds called alginate, to make semisolid gels, bind ingredients, and thicken liquids. Their timing was excellent: after the war, the new convenience-food companies were exploiting the latest food-preservation techniques and chemicals (think freeze-drying and sodium benzoate) and expanding their product lines. Out of the laboratories and test kitchens of Sara Lee, Betty Crocker, Hostess, Kraft, and Unilever came canned pie fillings, cake mixes, Twinkies, ice cream, processed meats (to hold the bits together), mayonnaise, salad dressings, artificial maple syrup, candies with liquid centers, toothpaste (instead of tooth powder), shampoos, conditioners, lotions, shaving cream, lipsticks, air fresheners, and many other products, all made possible (for better or worse) because of phycocolloids. The paper industry used tons of carrageenan and alginate as sizing to give their products a smooth surface, and textile printers found that phycocolloids prevented dyes from running (so a polka-dotted fabric, for example, did not become polka-blobbed). The pharmaceutical industry also used carrageenan and alginate as binders to hold pills together. Brewers used it to "fine" or clarify beer that otherwise had a haze of proteins and polyphenols. Scientists discovered they could add alginate to heartburn medicines — the alginate worked with bicarbonate to create a barrier so stomach acid wouldn't rise up the esophagus. Phycocolloids are still important in all these products and industries today.

The Algin Corporation, which became Marine Colloids, paid college students, children, and off-season lobstermen to gather Irish moss from the shores of Massachusetts, Maine, and the Canadian Maritime provinces. For many ocean-side residents, summertime meant rising before dawn and spending days clambering over rocks or standing in dories raking up hundreds of thousands of pounds of seaweed. Adult "mossers" could collect as much as a thousand pounds a day — if they returned at night to work the second low tide — and earn a year's college tuition in a summer. But there was money to be made even if you didn't want to get your feet wet or didn't have the back strength that mossing demanded. The company also hired people to spread the seaweed on the beach to dry, and then to pack it into crates to ship to the factories. In a town like Scituate, summer mossing was a cherished part of the culture, and an important source of income for the locals. (You can learn more at the town's charming Maritime and Mossing Museum, open only on Sunday afternoons.)

By the 1970s, Marine Colloids was the largest carrageenan manufacturer in the world, and North American *Chondrus* was no longer enough to supply its factories. The company sent Louis Deveau, one of the young employees at its Canadian subsidiary, to Mexico and East Asia to find new sources of seaweeds for the extraction plant in Maine. Although Irish moss doesn't grow in those warmer waters, Deveau discovered that another red seaweed, *Eucheuma cottonii,* was a fine substitute. With Marine Colloids and others eager to buy, Indonesian and Philippine fishermen learned to cultivate *cottonii,* growing it in quiet ocean bays on strings they stretched between stakes set into the sand.

When the American multinational FMC bought Marine Colloids in 1980, Deveau bought the Canadian subsidiary and named it Acadian Seaplants, Ltd., which we visited earlier in this section. He and his son Jean-Paul, who is now Acadian's CEO, expanded the company's focus and began a rockweed operation. North American livestock farmers had been importing rockweed from Europe, but the supply was unpredictable and customers were often left short. Acadian filled the market and more. The new line was a success, but, as Jean-Paul explained to me, "We recognized that the story of the product's efficacy was one hundred percent anecdotal. So, we invested heavily in science," setting up the research lab in Cornwallis and working with the National Research Council Canada.

The expansion was fortunate; by the early 1990s, Acadian's Irish moss business was melting away. The *cottonii* farmers in East Asia were supplying carrageenan-rich seaweed at prices no one in North America could match. Acadian exited the carrageenan business, leaving it to the Philippines and Indonesia.

Today, the carrageenan business is a mainstay of coastal villages in the Philippines, where small-scale family operations grow the bulk of the seaweed. In the twenty-first century, carrageenan farming has helped save tens of thousands of people from poverty. *Cottonii* farming is especially critical now, when foreign trawlers are illegally ransacking the offshore waters and driving local fishermen to despair. In desperate (and ultimately counterproductive) efforts to earn enough money to survive, fishermen are dynamiting coral reefs, a technique that kills not only salable fish but every other creature in the water and, worse, obliterates the corals the fish depend on. In the face of a drastic loss of fishing income, the carrageenan industry has become increasingly important. Entire families are involved in making the lines that seaweed is grown on, tying the seaweeds to the lines staked in shallow waters, and tending and harvesting the crop. The seaweed was once shipped abroad for transformation into carrageenan, but in recent years, processing has shifted to local factories, creating much-needed new jobs.

Unfortunately, the well-being of these communities is now threatened from a new source. While carrageenan has been widely used and approved for consumption for more than sixty years, some individuals have been claiming that it's unhealthy to eat. Public personalities such as Dr. Josh Axe, Dr. Andrew Weil, and Vani Hari ("Food Babe"), proponents of complementary and alternative medicine, assert via their websites and blogs that carrageenan is an inflammatory agent and a hazard. Never mind that the US FDA, the European Union, the UN's Joint FAO/WHO Expert Committee on Food Additives (JECFA), and the Japan Ministry of Labor and Welfare have determined that carrageenan is safe to consume.

How did this misconception arise? The concern over carrageenan was first expressed in an article by Dr. Joanne Tobacman, who studied a degraded form of carrageenan called poligeenan, and concluded it has harmful gastrointestinal effects. Poligeenan, however, is substantially different from carrageenan. It doesn't have any of the thickening, emulsifying, or other properties that make carrageenan so valued. In fact, it isn't used in foods at all, only in diagnostic medical applications. The JECFA considered Dr. Tobacman's evidence and concluded that carrageenan is safe, without limitations, for human consumption. Nonetheless, certain wellness gurus have gone on a campaign against carrageenan, and many US companies, responding to consumer concerns, have removed it from their products. This has done absolutely nothing for anyone's health, but it has reduced the market for carrageenan, and thereby the incomes of poor Filipino and Indonesian farmers.

I drink lactose-free, calcium-fortified milk; a minute quantity of carrageenan in the milk keeps the extra calcium suspended. I was delighted, when I started my research, to find I was consuming algae every day. I still am.

Some seaweeds produce a colloid that saves lives. The story starts in Japan, on a winter night in 1658, when a Japanese innkeeper named Mino Tarōzaemon tossed the remnants of his seaweed soup — made from a *Gracilaria* or *Gelidium* species — out the kitchen door. The next morning, he found the soup transformed: it had become a clear, pink-tinged solid. In that moment, he inadvertently discovered that certain seaweeds make phycocolloids that, under the right conditions, form a gel. (More technically, they form a chemically linked, transparent, semirigid, three-dimensional lattice.) Tarōzaemon and other Japanese cooks learned how to extract the phycocolloids without having to boot the soup and used them to make food, especially desserts. The substance in those seaweeds is now known as agar; the name comes from the Malay word for jelly, *agar-agar*.

Agar leapt from mild-mannered kitchen companion to lifesaving superhero in the late nineteenth century. In 1870, Louis Pasteur's germ theory of disease was still new and far from universally accepted. Pasteur had demonstrated that microbes (rather than "noxious miasmas") cause disease, and that they arise only from other microbes (rather than spontaneously popping into being). He didn't, however, develop experimental proof that one particular microbe causes one particular disease. That's where Robert Koch, a Prussian doctor working out of a makeshift lab in his apartment in the mid-1870s, stepped in. Koch set out to prove that the anthrax bacillus microbe causes anthrax, a major disease in cattle. He began by inoculating mice with blood taken from the spleens of animals who had died of the disease. Every time, the mice died. Still, this was not categorical proof that the bacillus was responsible; it was conceivable that some other substance in the blood had killed the mice.

Koch knew that the only way to provide definitive proof would be to inoculate an animal with a pure culture of the bacillus, thus establishing that the microorganism was the only possible source of disease. But at the time, creating a pure culture of any bacterium was an exceedingly difficult task. Cultivating it in a liquid nutrient broth didn't work, chiefly because the microorganisms dispersed and were difficult to collect and concentrate. Koch did finally manage to culture pure *Bacillus anthracis* by growing it in a drop of sterile fluid that he extracted from behind the lens of an ox's eye. But using bovine aqueous humor as a culture medium was hardly a practical lab technique.

It would be better, Koch realized, to grow bacteria on a solid nutritive substance so the microbes couldn't drift apart. He inoculated slices of potato and had some success: the microbes reproduced as a colony in a dense, pure mass. But potato slices were impossible to sterilize without reducing them to mush, and their opacity made it difficult to observe and measure the dimensions of bacterial colonies.

Next, Koch tried gelatin made from the collagen in the skin, bones, and connective tissue of animals. (A little nutrient broth had to be mixed in since bacteria have difficulty digesting collagen.) Gelatin as a culture medium had the advantage of being solid, clear, and sterilizable, but it also had a major disadvantage: it melts at 95 degrees Fahrenheit (35 degrees Celsius). Because Koch wanted to investigate microbes that thrive in mammals, and their normal body temperatures are generally above 95 degrees Fahrenheit, gelatin was impossible. The biologist was stymied.

Fortunately, he was not alone in trying to deal with this problem. Dr. Walther Hesse, a German researcher investigating the presence of microbes in air, encountered the same difficulty. It was his American wife, Fannie, who found an answer. Fannie, who doubled —or rather, tripled—as his lab assistant, medical illustrator, and the major-domo of their family of five, had a recipe, passed along from friends who had lived in East Asia, for making jams and jellies thickened with agar-agar. Frau Dr. Hesse knew her preserves didn't liquefy in the summer heat, so she suggested to Walter that he try the gel in the lab. Mixed with some nutrients, the medium, which remains solid up to 140 degrees Fahrenheit (60 degrees Celsius), proved ideal. Hesse immediately gave the recipe to Koch, who, in 1881, used it to isolate the bacteria that cause anthrax, tuberculosis, and cholera.*

Agar quickly became as important to microbiologists as the microscope, and it's just as critical as a growth medium in labs today. But agar has found new uses in forensics, pathology, and in paternity testing: the micron-size spaces in its tough matrix of polysaccharides act like the holes in a colander to separate and sort DNA and protein molecules in a commonplace process called agarose gel electrophoresis. More than 150 years after Koch first used it, agar still has a prominent place in medicine and science.

• • •

*An assistant in Koch's lab named Julius Petri invented a shallow, two-part glass container to hold the medium. The container, the ubiquitous Petri dish, is named after its inventor. But few remember Fannie Hesse, the woman responsible for the medium inside.

In the twenty-first century, most agar is primarily made of the macroalgae species *Gelidium*. These seaweeds have small shrublike thalli (bodies), and grow most abundantly eight to sixty feet below the surface of the cold, turbulent waters off the coasts of Spain, Portugal, and Morocco. Professional divers reap *Gelidium* from the seafloor using a kind of underwater mower that allows the seaweed to regenerate its blades. Loose seaweed also collects in underwater depressions, which harvesters suction into boats. Where strands wash up on beaches, collectors rake them up with tractors.

Unfortunately, in many areas, harvesting *Gelidium* has been an unregulated free-for-all, with predictable results. In 2010, the Moroccan government announced its *Gelidium* harvest had declined so much that the government was setting export quotas to protect the species' long-term survival. Prospects for new sources of *Gelidium* are limited. Because the seaweed only grows in highly agitated conditions — conditions not easily replicated in aquaculture — there is no farmed alternative. The price of agar has nearly tripled in the last seven years, and there has been talk of hoarding. There are other species — *Gracilaria,* in particular — that produce a lower-quality agar, but the shortage may soon have an impact on research and in diagnostic labs around the world. Who knows, we may be back to oxen's eyes.

Phycocolloids have been around for hundreds of years, but scientists are exploring new uses for them. Skin cancer is on the rise worldwide. Most sunscreen lotions are made of compounds that either absorb or physically block UV rays. While sunscreens do indeed protect skin, we now know that those containing oxybenzone and other common related compounds are deadly for marine life, even in small quantities. The problem is that between six thousand and fourteen thousand tons of the stuff wind up in the ocean each year, concentrated in areas that tourists visit — like coral reefs. Out of concern for the Hawaiian Islands' endangered reefs, in 2021, Hawaii will ban the sale of nonmineral sunscreens. Only those based on zinc and titanium oxides will be available for purchase in the state.

It is good news, then, that researchers in Sweden, Spain, and

Britain, in separate investigations, are finding that algae-based sunscreens are highly efficient at absorbing UVA and UVB radiation. Of course, this news should be no surprise: algae have been perfecting their all-natural, nontoxic, sun-blocking techniques for more than three billion years. And researchers at the College of Pharmacy at the University of Florida have bioengineered a cyanobacterium to produce much more of the specific molecule, shinorine, that provides the UV protection. No worries, by the way: shinorine is clear, not green.

New medical uses for phycocolloids are also in the offing. For example, scientists at NIH's National Institute for Biomedical Imaging and Bioengineering are developing an alternative to insulin injections for diabetics — a patch made of alginate. The patch, applied to the skin, is able to penetrate its outer layer and not only deliver insulin painlessly, but provide it steadily over time.

Scientists are also finding a new role for alginate in fighting microbial infections. Eighty percent of infectious bacteria form biofilms — slick, sticky aggregations of themselves — that allow their cells to communicate with one another to evade both the body's innate defense mechanisms and pharmaceutical antibiotics. Biofilms also enable bacteria to adhere to a surface — say, lung tissue — to keep an infection active and in place. They are a hallmark of cystic fibrosis, a genetic disease that causes the lungs and digestive system to become clogged with mucus, which then becomes a home for infectious microorganisms. For the last decade researchers at Cardiff University have been working with the Norwegian company AlgiPharma to develop an alginate drug that disrupts these membranes. This promising drug is currently in clinical trials with cystic fibrosis patients and could open the door to treating many other diseases.

The phycocolloids are old dogs, but they're learning new — and lifesaving — tricks.

3

Land Ho, Going Thrice

For years, Acadian was out of the Irish moss business, but the seaweed has made a comeback at the company. Resurrection came in 1992 at an international seaweed conference, when a Japanese food manufacturer, intrigued by *Chondrus crispus*'s flower-like form, told Deveau that he thought there would be a market for it as a salad ingredient or garnish in Japan. But, he added — and it was a huge "but" — only if Acadian could change its natural color. According to the Japanese tradition of *Washoku* or "harmony of food," he explained, an appetizing and healthy meal has five color elements: white and black — often provided by rice and nori — plus green, pink, and yellow. If Acadian could produce its Irish moss in those three colors (and, by the way, in uniformly shaped pieces with no blemishes), there was money to be made.

Today, Acadian produces more than two million wet pounds of Irish moss in three colors, and dries and ships all of it to Japan. The seaweed, packaged under the brand name Hana Tsunomata, is cultivated on land at a facility not far from Yarmouth. The process starts in the lab — off-limits to outsiders like me — with tiny bits of Irish moss snipped from a single mother plant and nurtured under grow lights. After a year, when the young seaweeds are about the size of a fist, they are transferred outdoors to one of dozens of shallow, lined "bubble ponds" filled with filtered seawater from the adjacent coast. Acadian's Jeff Hafting and I walked around the ponds, which appear to be on a low boil thanks to air pumped up into them from a network of pipes below. They cover more than seven acres and are the seaweeds' home for about six months.

We stop by a pond nearly ready for harvesting, and Hafting fishes in the water for one of the dark maroon pom-poms that

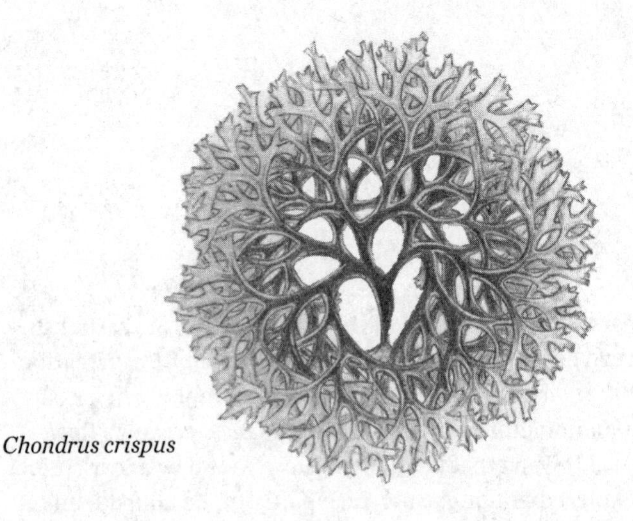

Chondrus crispus

tumble about, gently buffeted by the currents. In the wild, the seaweeds would attach to an underwater surface and grow in an upright, shrublike form, but here, in constant motion, they have no sense of up and down, so they grow equally and identically in all directions. When he pulls one out, it lies flat and is nearly as big as his outstretched hand. Its fronds look and feel as if they were made of thin, flexible, high-quality plastic, stamped out by machine.

It takes about eighteen months to grow a *Chondrus* seaweed to full size in these ideal conditions. Across the way, a truck pulls up to harvest the contents of a pond. What looks like a giant rectangular colander swings out from the bed of the truck and plunges into the water to scoop up the seaweeds. After the water drains out, the colander tucks itself back onto the truck, which motors off to one of the nearby buildings, where the seaweeds will be convinced to cloak themselves in more desirable hues.

Although the process details are a trade secret, the general approach is millions of years old: It's how the green leaves on deciduous trees become yellow, orange, and red in the fall. Chlorophyll, the pigment in leaves that gives them their green color, gradually degrades as temperatures fall and the days grow shorter, allowing other, hidden pigments in the leaves their moment to shine. The scientists at Acadian have developed a process that degrades the seaweed's dominant phycoerythrin and other subsidiary pigments so that the desired colors emerge. Once transformed, the seaweeds go into silo-like dryers, where they fly through the air, bumping against each other like popcorn in a movie theater popper. They knock any foreign substances off their surfaces and break themselves into the uniform florets that Japanese customers demand.

As I leave, Hafting gives me a bag of Hana Tsunomata, which contains a mix of all three colors. I have to say, the contents of the bag don't look promising; they could pass for crumbles of dried parsley, except in deep maroon and ochre, as well as dark green. Back at my hotel, I fill a water glass and drop in a teaspoon of the crumbles, and they immediately start to unfold and expand. It's as if I'm watching a time-lapse film of flowers blooming: in five minutes, I'm looking at a delicate, floating, spring bouquet of florets in new-leaf green, cherry-blossom pink, and marigold yellow. I pluck out a piece and nibble it. It's slightly crunchy but has no taste; the pleasure is all in the presentation. At home, as the package suggests, I use them as a striking garnish and for brilliant highlights in a spring salad.

Acadian isn't out to change the world by growing beautiful Irish moss. But I'm intrigued by the idea of growing seaweed on land. If wild harvesting was the first step in exploiting seaweed for food, and growing it as a crop in the ocean was the second, could growing seaweed on land be the third?

Seakura, a young Israeli company, is exploring just that. A chief selling point for eating algae is that it's full of the nutrients it takes in from seawater. But while this tendency is a blessing when the seawater is clean, that benefit becomes a liability when it's not. Not every country has — or will continue to have — waters as pristine as South Korea's Jeolla province or the Gulf of Maine. And, of course, for landlocked countries, provinces, or states, seaweed cultivation has never been an option.

Yossi Karta, the founder of Seakura, an aquaculture company

located near the Mediterranean Sea, just north of Tel Aviv, has developed new technology that he hopes will vastly increase the amount of clean seaweed available to the world. The company cultivates its crops in circular tanks housed in what are essentially plastic greenhouses. The structures allow light to pass through while excluding airborne pollutants. Although the tanks are filled with seawater, the water is drawn not from the Mediterranean but from deep marine wells drilled into the limestone that underlies the land in this area. Seakura is fortunate: not only does the limestone filter pollutants out of the water, it also enriches it with calcium and potassium.

In a company video, I watch thin sheets of green *Ulva* and red tumbleweeds of *Gracilaria* swirl around the tanks, impelled by a gentle current. In this setting, controlled to deliver the ideal nutrients, light, and other growth factors, the seaweeds grow to full size in only five weeks, and Seakura is able to harvest as many as nine crops per year. That means the tanks produce one hundred tons of seaweed per acre, far more biomass than could be harvested from an acre of coastal seawater. The seaweeds are high quality: they are about one-quarter protein (including all nine essential amino acids), 50 percent fiber, and have a good array of minerals and vitamins, and they have no mercury, arsenic, or any of the other heavy metals that seaweeds can pick up in the wild. Seakura sold most of last year's crop in Israel, with some sales to British and Belgian stores. Most of its product is sold as a powder and flakes, but the company also sells pasteurized seaweed purée and fresh frozen cubes to add to smoothies or stews. A serving of purée is advertised as having as much iodine as a serving of haddock, as much manganese and magnesium as an equivalent serving of kale or spinach, and as much fiber as an apple.

The largest potential market for Seakura's seaweed is food manufacturers looking for a natural way to enhance products like soups, crackers, dressings, and snacks with the vitamins and minerals that seaweeds provide. When I spoke to Karta by phone and asked him about his company's goals, he told me that, although he plans to expand his Israeli facility, his primary goal is to license Seakura's technology as widely as possible.

Karta is not your average entrepreneur. He is a fervent prophet of immaculate algae, which he sees as the God-created basis for all life on land and critical to the evolution of our species. To him, seaweed is the original staff of life, and he is committed to seeing that mankind — including people who live far from coasts — have access to the health benefits of an untainted source. (In this respect, Karta reminds me of John Harvey Kellogg, the physician and founder of the Kellogg's food company, whose plan to "harmonize science and the Bible" led him to become an evangelist for the whole-grain breakfast cereals he and his brother invented.) Karta is distressed — truly, to the core of his being — by the quality of some of the seaweeds on the market. His goal is to see land-based seaweed aquaculture, using filtered seawater or artificial seawater created from a clean municipal source, prosper around the globe.

Karta mentions that Seakura consults from time to time with Dr. Yonathan Zohar, chair of the Department of Marine Biotechnology at the Institute of Marine and Environmental Technology at the University of Maryland, who focuses on sustainable, land-based fish aquaculture. Zohar raises fish in tanks similar to Seakura's, he tells me, and is also an expert in growing microalgae to feed aquacultured fish. He and Zohar have discussed how Seakura's seaweeds might be used as feed for aquacultured fish. Of course, I find this intriguing, and since Zohar's office is less than an hour away in an office building in downtown Baltimore, I call him up to arrange a visit.

The man who greets me in his office is tall, slender, and white-haired and speaks in a resonant voice with a soft Israeli accent. We head down a flight of stairs and then through a heavy metal door. I'm imagining a couple of quiet, brightly lit rooms and the usual lab setup: benches, hoods, microscopes, flasks, shaker plates, lots of analytic instruments and other equipment, everything brightly lit, white, and sterile. Instead, we enter a dim, noisy, half-acre space with low ceilings that is filled with what look like pale green backyard swimming pools. Pipes and ductwork for the building's mechanical systems run overhead, fluorescent shop fixtures hang from the ceiling, and there's a hum of machinery in the background. What I'm looking at is an indoor fish farm in a downtown office building.

We go up to one of the tanks and look in. Swimming swiftly against a current are several thousand silvery, six-inch branzino, also known as European sea bass. Zohar tosses in a handful of fish feed and they swarm to the top, roiling the surface. Other tanks we look at contain schools of more mature sea bream, dorado, rockfish, and other popular table fish. A few smaller tanks, partially enclosed (for privacy?), contain full-grown, slow-moving, breeding stock that hover in place or swim lazily around. Much of Zohar's research has been on the conditions that induce spawning and how to stimulate different species to reproduce in tanks. In this sunless lab, Zohar and his colleagues and students raise two tons of perfectly healthy fish per year.

Best of all, the fish are grown in a way that has zero impact on the environment. The artificial seawater in the tanks — city water with a few common, inexpensive compounds added — stays crystalline and pure for the many months it takes to raise a market-size fish, thanks to two filtering devices that use microorganisms to constantly remove fish excreta. Zohar's fish are raised in pristine, controlled waters that emulate a species' wild environment. As a result, they are healthier and don't need antibiotics, and no fish waste leaves the building. (The fish themselves, however, eventually depart for Baltimore restaurants.)

The contrast with traditional outdoor aquaculture couldn't be more stark. When fish are farmed in the ocean, they are raised in floating net pens, and their waste becomes an environmental hazard. The excreted nitrogen and phosphorus spur the growth of algae and, consequently, respiring bacteria that reduce oxygen levels in the water, which puts stress on the fish.

They face other stresses, too. Unlike their wild brethren that range about from season to season seeking the best circumstances for growth and health, farmed fish live in one spot under, ipso facto, less than ideal conditions. They are more susceptible to parasitic diseases like sea lice, which can sicken or kill them outright.*

Some penned fish are now fed a less-than-ideal diet. Traditionally, as carnivorous fish in aquaculture grow larger, farmers have fed them a mixture of fishmeal and fish oil. Both are made from certain small, oily, bony, and less appetizing (to humans) "forage fish," such as anchovies, sardines, menhaden, and herrings, that farmed fish would have eaten in the wild. However, because forage fish are becoming scarce and more expensive, aquaculturists have been substituting fishmeal with soy and other plant proteins. It's no surprise that terrestrial plants are not perfect foods for fish: no fish ever evolved in a landscape of soybeans. Moreover, plants don't produce certain amino acids — lysine, methionine, threonine, and tryptophan — that fish need. So, while farmed fish are vital to the world's food supply, there is room for improvement in how they are raised.

Zohar and his team are working on that. They feed their fish larvae and young fish with the specific microalgae or microalgae-fed zooplankton they favor. The lab's "algae kitchen" — an array of floor-to-ceiling plastic sleeves lit by grow lights so that they glow in shades of brown to gold to green — grows a variety of species to accommodate the various fish on site. The microalgae contain all requisite amino acids, and some also have taurine, a related compound that helps fish maintain a balance of fluids, electrolytes, and calcium while also stabilizing their cell membranes. Microalgae are also added to the feed of larger fish to supplement a plant-based diet. It's a way of getting the value of the forage fish without taking forage fish from the oceans.

But Zohar also hopes to use seaweeds like those that Seakura grows in fish farming. While the large fish we generally prefer to

---

*In fact, according to a recent study and model published in *Marine Resource Economics,* sea lice currently cost salmon farmers about 9 percent of revenues. Dr. Ian Bricknell, professor of aquaculture at the University of Maine, estimates that mortality losses are about $525 million per year. In addition, the high level of infestation leads to use of antibiotics and chemicals. In Chile, where the farmed salmon industry has grown quickly, farmers used 1.2 million pounds of antibiotics in 2014, according to government and industry sources.

eat are carnivores, there are a few tasty species that eat macroalgae. One promising prospect is a species of rabbitfish, already a popular, high-priced menu item in East Asia. (If you're picturing fish with long ears and a twitching nose, I'm sorry. Their name comes from their vegetarian diet.) Not only are rabbitfish prized for their delicate and mild flavor, they have no bones and are rich in polyunsaturated fatty acids. They are generally wild-caught, but some are aquacultured in pens in East Asia. Zohar believes it is possible to raise them using land-based aquaculture and feed them with tank-grown seaweeds. If he succeeds, we will have a new, nutritious seafood source that uses no forage fish and whose culture has no adverse impact on oceans.

Inland fish aquaculture is spreading. In Vancouver, Canada, the 'Namgis First Nation runs a $7.6 million business called Kuterra that raises salmon on land in tanks. In Maine, Nordic Aquafarms is building a land-based salmon farm, and the Norwegian company Atlantic Sapphire has raised more than $100 million to build one in Florida. Land-based shrimp aquaculture is another growing industry; as of 2015, there were eleven companies in Indiana alone growing shrimp in tanks, and there are others in Maryland and Massachusetts. Shrimp also eat seaweed, and researchers have been exploring raising the crustaceans in the same tanks with sea lettuce. The results are promising: When the two were raised together, the shrimp had a better fatty-acid profile and higher carotenoid levels. Plus, the seaweed also helps clean the water by recycling the nutrients in fish waste.

At the moment, land-based fish aquaculture is limited by the high capital costs of establishing an operation; tanks, controls, and waste management equipment are expensive. However, consumer demand for healthy, sustainably harvested, and environmentally sound fish is growing, and with it, demand for high-quality fish feed will grow, too. Microalgae and seaweed, intensively cultivated in pristine conditions and entirely nonpolluting, can help fill that demand.

4

Seaweed Stuff

Seaweed has long been a source of food for animals and man, but it has a history — and perhaps a future — as a durable material. In the late 1600s, the Scots discovered that if you burn kelp, the remaining ashes are rich in two alkali compounds: soda ash (sodium carbonate) and potash (potassium carbonate). These two compounds, along with sand, are the major ingredients of glass.

Traditionally, glass had been made from the sodium carbonate of burned wood, but by the seventeenth century much of the Scottish landscape had been stripped of its trees for fuel and construction. While Britain had discovered it had plenty of coal that could be burned for fuel, coal didn't leave behind ash useful for glassmaking, so British glassmakers had to look outside the country for a source. On the continent, glassmakers used ash made from burning Mediterranean barilla, a type of marsh plant that accumulates high levels of salt. Barilla soda ash was a fine substitute for wood ash, especially since it has a high portion of sodium carbonate, which makes it particularly transparent.

The problem was that barilla soda ash was expensive and became ever more so during the 1700s as demand for fine glass grew in an industrializing and increasingly prosperous Europe. To make matters worse for British manufacturers, whenever England and Spain were at war, which they were for much of the eighteenth century, barilla ash was hard to come by. Fortunately, hard-pressed British glassmakers discovered that kelp ash could substitute for barilla. The resulting glass, blown into globes, flattened, and cut into small panes, was of low quality but it was cheap. If not beautiful, it served most people's needs.

Large-scale production of kelp ash began in Orkney in 1722. At the time, with the price of wool rising, landlords (lairds) were evicting smallholder tenant farmers in order to turn the fields over to grazing sheep. While some tenant farmers were forced to emigrate, landlords relocated others to small coastal tracts of land called crofts. The crofts were often unsuitable for grazing and barely able to support a farming family. The crofters had to turn to kelping – gathering seaweed and burning it – either because it was a condition of croft tenure or because it was the only source of income. The crofters resisted at first. "Their forefathers had never thought of making kelp, and it would appear that they themselves had no wish to render their posterity wiser in this manner," one observer wrote. The Orcadians hated the business and the oily black smoke it produced, fearing it would poison their land and animals and make their wives infertile. Nonetheless, as the price of kelp ash rose, the cottage industry spread throughout the islands and to the coast of the mainland.

The work was seasonal and hard. After the frequent winter storms, people would rake great drifts of seaweed, known as brook, that washed ashore, hurrying to capture it before it washed back out to sea. When the seaweed came ashore, elaborate rules governed who could harvest it and how much they could gather. Lots were drawn to determine which part of the drift a person was entitled to: seaweed in the middle of the brook, less tumbled about, was more valuable. Kelping, Samuel Johnson reported from a trip he took to the Hebrides with James Boswell in 1773, "excited a long and bitter litigation between the Macdonald and Macleod for a ledge of rocks, which, till the value of kelp was known, neither of them desired the reputation of possessing." In some places, seaweed was almost farmed. On sandy shores, locals would move rocks into the water to give the seaweed a foothold. Once the "sea-wrack" was gathered and dried, it was hauled by horse and cart along "wrack roads." The blades, which decayed too rapidly, were set aside for manure. Dried stipes were stacked and burned with heather and straw for four to eight hours in stone-lined pits and kilns built near the shoreline. When the ash built up to about eighteen inches, men raked it, then pounded it into a mass with iron clubs.

At a time of harrowing economic change, as many as a hundred thousand Highland Scots found work in kelping. Unfortunately, they could only sell their ash to their landlords, who paid as little for it as possible while squeezing their poor tenants with rent increases. The crofters were infuriated and powerless, but the lairds profited well, selling the ash to the Glasgow glass and soap factories. The laird of Ulva in the Hebrides, according to *The Memoirs of Sir Walter Scott*, "at once trebled his income and doubled his population . . . by dint of minute attention to his property and particularly to his kelp." Still, for decades kelping kept many a Scots family alive. By the early 1800s, Highlanders were gathering about 400,000 tons of seaweed, which meant 20,000 tons of ash, per year. In the Orkney Islands alone, 20,000 people — almost the entire population — were employed for a part of the year in kelping.

For a century, Highlanders survived and the lairds prospered from kelp ash. But by the late 1700s, only the high duties on Spanish barilla soda sustained its market price. When the wars with Spain finally ended in 1815 and the government lowered the tariff on barilla, the price of kelp ash dropped, sinking from £22 per ton in 1810 to £4 in 1830. The Scottish industry was devastated. Crofts became economically untenable, especially because the rented land had been subdivided among at least two generations of the original tenants' descendants. Evictions, starvation, and massive emigration followed. The kelp ash industry nearly disappeared.

But not quite. For a number of years, some remaining crofters made money on burning kelp for iodine to be used as an antiseptic. But then that industry vanished, too, when mineral deposits of iodine were discovered and exploited in Chile during World War I.

While algae are no longer useful in the glass and iodine industries, they have found their way into other products. Diatomaceous earth, the silica-rich fossil bodies of diatoms, is added to wall paint to control its reflectiveness. It's a slight abrasive, so you can find it in toothpaste and metal polishes, too. In 1867, Alfred Nobel added it to nitroglycerine to make dynamite, and diatomaceous earth is still used in the explosive today. And algae have a new and potentially far larger use, in a product that no one in the eighteenth century could have dreamed of.

In August of 2017, I saw an ad for a waterproof running shoe called Ultra III Bloom made by the UK company Vivobarefoot. According to the ad, both the uppers and soles of the lightweight, flexible shoes were made of algae. I suspected the amount of algae in the shoes would be de minimis, a mere marketing ploy, but I decided to investigate. Vivobarefoot directed me to Algix, the company that makes the shoe's material, and I called up Ryan Hunt, cofounder and chief technology officer.

Algae shoes, I was surprised to discover, are the real deal. About 50 percent of the plastic polymer in a Vivobarefoot shoe is made of algae. In 2018, the company's first full year of operation, it will use some twelve million pounds of algae — genuine pond scum — to make the foam components of Vivobarefoot, Altra, and Keen sport shoes, as well as Billabong and Firewire surfboards and Surftech standup paddle boards. This definitely needed further investigation, so I asked Ryan if I could visit.

Five days later, after a two-hour drive from the Birmingham, Alabama, airport, I'm waiting in the spartan lobby of Algix's factory in Meridian, Mississippi. Ryan is running a little late, which gives me time to look in the two display cases. The glass shelves hold samples of the company's Bloom shoe products: thick white midsoles; colorful, thin, and flexible insoles; and stiffer outsoles with their characteristic deeply incised treads. Stamped into the samples are familiar names — Keen, Clark's, UnderArmour, and Skechers — among a dozen others. There are also two complete shoes: the Vivobarefoot shoe and a slide shoe similar to a Croc.

I spot a man fumbling with his entry card at the glass lobby door. At a jazz club, I'd peg him for the saxophonist — dark hair standing on end, black glasses, black goatee — but this is Ryan. He doesn't look like a man whose career is in plastics.

Say *plastic,* and people think chemical, artificial, and unnatural. But plastics themselves are, molecularly speaking, organic; they're made of carbon-based polymers — that is, long chains of carbon combined with oxygen, nitrogen, or a few other elements, depending on the type of plastic. While most plastics today are made of ancient carbon that was produced by dead organisms eons ago and compressed over time into natural gas and petroleum, it's perfectly possible to make plastics with carbon that comes from organisms that died only very recently. Which is how Ryan got into the plastics business.

Ryan came to his career by accident. It was 2009, and he'd been working on his PhD in bioengineering at the University of Georgia, focusing on algae biofuel. "At one point, we were looking for assistance on a project that involved blasting algae with high-energy electromagnetic pulses to rupture the cell walls and extract oil. This guy named Mike van Drunen joined us; he'd done work on the exact same technology in the 1990s to sterilize beer and juice. The two of us were working away when one day a professor from the materials science department visited the lab, and was telling us about how he was using feathers and other waste chicken parts, compressing and molding them to make plastics.

"*Well, that's interesting,* I thought, *I have a whole bunch of this leftover algae.* 'See what you can do with this,' I told him. So, he took the algae and came back with a chunk of plastic.

"When Mike saw that, his eyes lit up. Well, it turns out that while I thought of him as an electrical nerd like me, he was also an entrepreneur-slash-plastics engineer who had started a pretty big plastic-packaging company."

Ryan and Mike realized they could do something, maybe something big, with algae plastics. Like chicken feathers, algae have a high protein content, and proteins—like plastics—are carbon-based polymers. In 2010, Mike and Ryan formed Algix and spent the next five years trying to understand how to turn whole algae into plastic, and then how to turn that plastic into marketable products.

One of the plastics they made is a version of ethylene-vinyl acetate (EVA), a polymer normally made from petroleum and natural gas. EVA is the feedstock for a wide variety of consumer goods, ranging from food wrap to soccer cleats to swimming pool noodles. Algix was able to manufacture an EVA that replaces half the fossil hydrocarbons in the plastic with algae protein polymers. At first, Mike and Ryan thought they'd manufacture and sell their Earth-friendly EVA pellets to downstream product manufacturers, just as Exxon does. But the would-be entrepreneurs soon realized that the business of making pellets is one of very large volumes and very small margins, and that they would never be competitive with the large-scale manufacturers who dominate the industry. Instead, they needed to be in one or more of those downstream businesses, some of which produce higher-value (and higher-margin) goods.

After intense experimentation, they developed filaments for 3D printing, biodegradable flower pots, containers, packaging, thin films and wraps, and several other possible products. The flowerpots were a technical success, but they were just a little more expensive to produce than a standard pot. While retail customers might pay a higher price for an environmentally friendly pot, the major buyers of flowerpots are horticulture companies, which would not. The films and containers had a similar problem. When consumers shop for these inexpensive goods, many choose the brand on the basis of price. An Algix food wrap would be a little more expensive than its competitors. Despite its positive environmental story, it would likely not be a winner. Of the first group of algae products, only the 3D printing business turned out to be successful.

Then, Algix realized that while making EVA pellets wouldn't be profitable, EVA foam — made by shooting molten EVA full of air to created trapped bubbles — might be. For forty years, the soles of running and other sports shoes have been made of such a foam, which is poured while still molten into shoe molds or spread out in a sheet and cut for insoles. EVA foam goes into the ten billion pairs of sports shoes manufactured each year, putting the bounce in our steps, cradling our feet, cushioning our joints, and supporting our arches. It's also incorporated into surfboards, sealants, helmets, floor mats, and dozens of other products.

Ryan tells me that Bloom foam — made from a fifty-fifty blend of algae EVA and fossil fuel EVA — makes not just a substitute sole, but a better sole. The algae polymers align and link with the petroleum polymers, strengthening the product and making it less likely to tear. They also disperse air bubbles in the foam more uniformly than petroleum-only polymers do, which makes Bloom EVA just a little more elastic. The algae component also confers two more modest advantages. For one, it modifies the foam's surface chemistry of the insole in such a way that the fabric covering on top of it is less likely to peel off. And, as a sole is manufactured and the shoe company's logo is cut into the hot foam, the logo retains a sharper outline.

For a while, Ryan and his business partners worried about the color of their product. Algix can produce foam in a rainbow of hues, but the chlorophyll in the algae gives them more subdued tones. But, it turned out that shoe manufacturers are content with the options; the colors signal that shoes made with Bloom foam are more natural, a selling point for environmentally conscious runners.

The manufacture of sports shoes is notoriously hard on the environment. They are made almost entirely of petrochemicals, and most are manufactured in Asia, where factories run on electricity generated by coal-fired power plants. The result, according to research from MIT, is that producing one pair adds thirty pounds of carbon dioxide to the atmosphere, the equivalent of keeping a 100-watt light bulb on for a week. Not so with the Vivobarefoot's algae shoe. The "eco-facts" of its manufacture are incised on its inner sole: 57 gallons of water cleaned and 40 balloons' worth of CO_2 prevented from entering the atmosphere. Wear algae shoes, and you leave behind a lighter carbon footprint.*

The raw material of Bloom foam is ordinary algal blooms. You might think — as I did — that nothing could be easier to procure than pond scum, but finding enough of it, Ryan tells me, is his chief

*Disposal of the shoes remains a problem; they end up in landfills because EVA is not generally recycled. Still, the algae portion of the shoes is carbon-neutral, having taken carbon dioxide out of the atmosphere, and in a landfill that carbon — as well as the petroleum-based carbon — may be sequestered from the atmosphere for up to a thousand years.

headache. "Back in 2009 and 2010, there was a lot of passion and promise in the algae fuel sector, and we expected that algae fuel companies would be leaving a lot of waste. But, it didn't turn out that way." At first, Ryan looked to municipal water treatment facilities, mostly in California, that send treated, clean water through a tertiary cleaning process. In this last cleaning, water goes into ponds or indoor pools where microalgae are grown to absorb nitrogen and phosphorus. The dead microalgae, which are usually discarded in landfills, are perfect for Algix. The problem is that each facility only produces a couple thousand tons of algae per year, and Algix needs far more.

"In 2012, we were so desperate for algae," Ryan told me, "I thought we were going to have to build our own raceway ponds, but, of course, we didn't have millions of dollars to do that. So, Mike and I went scouting on Google Earth, looking for bodies of water covered in the green of algal blooms. When we looked at Mississippi and Alabama, our eyes were like giant gumballs." What they had discovered were the states' 96,000 acres of catfish farms. Bingo: the ponds are full of algae.

Catfish farmers feed their fish with fishmeal. There's no spoon-feeding a catfish, which means that uneaten feed ends up nourishing algae. For every pound of fish the farmers grow, they also grow about a pound of algae, which has to be disposed of. So, when Ryan asked the farmers if they would sell him their pond scum, they were delighted. Algix established its factory outside Meridian for its proximity to catfish farms, as well as the Gulf ports.

It's time for us to head into the Algix factory, a spotless and brightly lit two-acre space with a corrugated metal roof that rises several stories overhead. Our first stop is one of the mobile harvesters, which happens to be in for repair. It's not much to look at: a flatbed trailer topped with a steel box the size of a Mini Cooper, various pumps and white pipes, and a large white bin. When the harvester is operating, it sucks in several hundred gallons of pond water per minute, concentrates the algae, and then sends a stream of thick, lumpy, grass-green slurry—about a thousand gallons a day—into the bin at the back end. (Don't worry about the catfish; a screen over the intake pipe prevents them from becoming purée.) Meanwhile, the cleaned and oxygenated water is piped back into the pond.

An Algix tanker truck then hauls the slurry to the factory. Ryan shows me where it drops into a hopper and then is spread out on a conveyor belt. The belt enters a line of eight industrial microwaves that zap the algae with a total of a million watts of power, vaporizing the water inside them and exploding their cells. The vapor is evacuated through broad silvery ducts that exit through the roof, while the particles of dry algae continue into a high-speed jet mill where, smashed into each other at the speed of sound, they are pulverized into a fine powder.

The powder then travels by vacuum tubes to the compounding complex, the repository of Algix's proprietary technology. The complex looks to me like an elaborate three-story jungle gym painted yellow, white, blue, and silver. In here, the algae powder, pellets of ExxonMobil's plastic resin, and various additives (as specified by each customer) are melted and fed into an extruder, a meters-long set of rotating twin screws made of dozens of custom-designed and machined parts. At the far end of the extruder emerges a thin, green-brown, half-algae, half–fossil fuel plastic strand, which then travels into another machine to be chopped into small pellets. Finally, the pellets are shipped to contractors in China who turn them into finished products. In 2018, the company acquired twelve million pounds of algae for its current orders. That, Ryan believes, is just the beginning.

Algix has a dozen algae harvesters but will need dozens more to secure enough local algae for its orders. At a cost of $1 million per unit, however, the company doesn't have the capital to have more manufactured right now. Instead, it's found millions of pounds of algae already harvested and partially dried in China.

Lake Taihu, which lies seventy-five miles west of Shanghai, is China's third largest freshwater lake. Once an idyllic vacation destination, in the past several decades the surrounding region has become home to millions of people, as well as chemical and textile plants and farms. Industrial wastewater, sewage, and fertilizers now pollute the lake, and every summer a thick layer of vivid green algae spreads across a third of its nine-hundred-square-mile surface. When the algae die and decompose, the stench is nauseating. A few years ago, two million people in the surrounding area lost access to drinking water during a particularly bad bloom. (When people opened their faucets, green slime flowed into their sinks.) Although the government is trying to limit the inflow of pollutants, change is slow in coming. Fortunately, however, the prevailing winds tend to push algae into the northwest corner of the lake, which acts as a natural catchment, and as a stopgap measure, the government built sixteen shore-based harvesting stations to pull algae from the water. Each station produces about three thousand tons of partially dried algae per year.

In 2016, Algix negotiated an agreement with Chinese authorities that allows it to buy Lake Taihu's algae. The Chinese blooms now provide the bulk of Algix's feedstock, but Ryan hopes to gather more algae closer to home. While the low price of the Chinese product, including shipping, makes his Earth-friendly soles economically viable, he is chagrined by the environmental cost of shipping algae halfway around the globe. To use more local algae, Ryan tells me, Algix needs more — and more efficient — harvesters. The necessary research and acquisition, however, will have to wait until the company generates more revenue.

Plastics manufacture is a sophisticated business with many thousands of different formulations and products. Algix has tried substituting algae in only a relative few; its engineers are now working on the higher-value thermoset plastics used in certain pipes, epoxy, and seamless flooring. But algae can't be incorporated in all products. Some plastics require particularly high heat to make, but if algae are heated above 482 degrees Fahrenheit (250 degrees Celsius), their polymer chains break apart. Because the tough polymers in polycarbonates and nylon must be heated well above that temperature, we'll never have algae car bumpers or algae toothbrush handles. On the other hand, Styrofoam plastic is formed at relatively low temperatures, so packing peanuts and coolers made from a mix of algae and petroleum are technically possible, although their low market price is an obstacle. Still, given the three hundred million pounds of plastics produced around the world each year, there should be no shortage of market opportunities.

The manufacture of traditional plastics sends greenhouse gasses into the air, but the plastic objects are themselves an environmental problem. Few bacteria can degrade them, and that means our discarded grocery bags, fishing nets, milk jugs, clamshell packaging, and thousands of other items will linger for hundreds of years after we've discarded them. Less than 10 percent of the plastics manufactured in a year are recycled, and that percentage is bound to fall now that China is no longer accepting Western recyclables.

Between five and twelve million metric tons of plastic already make their way into the oceans annually. There, wind and waves break them into small pieces—some no larger than a cyanobacterium—that drift around in ocean currents. These microplastics look like food to zooplankton, fish, seabirds, whales, and the other marine animals that consume them. Some of these creatures die from toxins in the plastic; some are killed when the particles block their digestive systems. Others survive and are eaten by larger animals, which means the plastics and toxins work their way up the food chain . . . to us.

If only plastics weren't so damn durable. For many years now, companies have been making biodegradable plastics from plants—you may have used packing peanuts made from cornstarch instead of Styrofoam. But plants require arable land and fresh water, so plant-based packing materials don't really make much sense for the environment. Taking a different approach, scientists are now able to harness certain bacteria that naturally—or with the help of genetic engineering—synthesize plastic polymers. These bacteria have the potential to produce tons of biodegradable plastics. There's a catch, though: Because the microorganisms are heterotrophic, they need to feed on sugars. Not only does that raise the question of arable land and fresh water again, it also makes bacteria-created plastics too expensive to compete with petroleum-based plastics.

There may be a way out of the cost conundrum, though, and it depends on cyanobacteria. In 2017, Daniel Ducat, Taylor Weiss, and Eric C. Young, researchers at Michigan State University's Plant Research Laboratory, genetically tweaked cyanobacteria so that they constantly leak some of their sugars (which they produce, of course, through photosynthesis). When the researchers put the leaky cyanobacteria in the same tanks as the plastic-producing bacteria, they created a dynamic duo. The cyanos feed the bacteria, which then make plastics. In tandem, the two make a biomass that, after processing, is nearly 30 percent plastic polymer. Moreover, the algae-powered bacteria make the plastic twenty times faster than the bacteria alone can, which means twenty times more plastic and no expensive sugar. This could be a game-changer.

The prospect for more algae plastic is especially important because worldwide plastic production is accelerating; researchers at the University of California, Santa Barbara, led by Dr. Roland Geyer, noted in a 2017 study that half of all the plastics ever made — 8.3 billion tons — had been made in the prior thirteen years. Exxon predicts that 20 percent of the world's fossil fuels will be used to make plastics by 2050, double the amount today. If we were able to substitute a half-algae plastic for standard plastic, we could rein in future greenhouse gas emissions.

In a way, we're back in 1869, just before John Wesley Hyatt invented the first plastic. The country was in the midst of a billiards craze — there were eight hundred pool halls in Chicago alone — and the demand for ivory for billiard balls had skyrocketed. Hunters in Africa and South Asia were killing elephants by the thousands, putting the animals, as the *New York Times* put it, in danger of being "numbered with extinct species." Billiards manufacturers could see that they too might be headed for extinction — neither wood nor any metal made a satisfactory substitute — and one firm offered a $10,000 award to the person who could come up with an ivory replacement. Hyatt, a young printer in New York, combined cotton, nitric acid, and solvents to create celluloid, the world's first synthetic material. Plastic saved the elephants.

Algix had only been operating for two years when I finished this book, but in 2019 Adidas, Clarks, and Toms Shoes ordered algae soles for some of its models. The company is now sourcing its algae from water treatment facilities in Utah and Wisconsin, algae blooms in Florida, and algae-based omega-3 oil producers in New Mexico and Texas. The Michigan State Research is even newer. It's too early to know whether or not algae-based plastics will be the new celluloid and change the future of manmade materials. But it's clear that this time a new material would help us save ourselves.

5

Algae Oil

The prospect for algae plastics is intriguing, but, at present, manufacturing plastics is responsible for only 10 percent of our total greenhouse gas emissions. An algae substitute for transportation fuel would have a far greater impact on fossil fuel use and climate change. According to the Energy Information Administration, the transport of people and goods accounts for about 25 percent of all energy consumption in the world. Which is why, back in 2008, when I visited Valcent Products, the biofuel company outside El Paso, Texas, that was trying to make oil from algae in a greenhouse, I was inspired to write this book. Valcent founder Glen Kertz claimed he could produce 100,000 gallons of algae fuel per acre. That meant that if we devoted a million acres of the American Southwest's semiarid land (the government owns tens of millions of acres in Nevada alone) to algae, we could replace most of the 127 billion gallons of gasoline, diesel, and jet fuel Americans burned in 2015.

That, I thought, was an exciting proposition, and I was not alone in my enthusiasm. Major news outlets—Bloomberg, ABC News, Fox, and many others—visited Valcent, broadcast footage of the glowing green panels of algae, and reported the promise of this new fuel. And Kertz was far from the only entrepreneur on the "oilgae" bandwagon. In 2009, for instance, CEO Robert Walsh of Aurora Biofuels, speaking on PBS's *Nova,* said his company was on its way to producing 120 million gallons of fuel per year.

It was easy to be seduced by Valcent and its fellow algae oil startups. For one thing, the basic concept behind these companies made sense: Petroleum is dead algae crushed underground for millions of years. So, why not speed up the process and make oil from algae living today? Instead of pumping fossil oil out of the ground and adding carbon dioxide *to* the atmosphere, we'd be growing algae and pulling carbon dioxide *from* the air.

The promise of algae oil starts with the fact that all organisms convert some of the sugars they either make, in the case of algae, or eat, in the case of animals, into lipids that contain more than twice as much energy per gram as sugars do. They're a sort of magic suitcase, one that stays the same size even when you pack twice as much food in it. If you're an alga stuck in a shady spot where you can't photosynthesize, you can survive with the lipids in your go-bag.

Lipids come in a number of forms: We humans stash our emergency energy as solid fats, sperm whales use waxes, and algae store theirs as liquid oil. While the amount of oil that algae socks away varies among species, some are particularly prudent and store half their weight in oil. There are even a few hypercautious types that, when they sense hard times ahead, will bank as much as 85 percent.

When fossil fuels were cheap and plentiful, as they were until the middle of the twentieth century, no one thought about alternative energy sources. But during World War II, the military's extraordinary demand for fuel overseas led the US government to ration it at home, and a few scientists began to consider algae oil as a possible substitute. After the war, however, when petroleum supplies rebounded, that interest evaporated. Then, in 1973, the Organization of Petroleum Exporting Countries (OPEC) imposed an oil embargo against the US for having resupplied Israel's military during the Arab-Israeli War. Americans faced long lines at gas stations and skyrocketing fuel prices; the US recognized the dangers of its dependence on foreign fuel stocks. The embargo ended in 1974, but the West's vulnerability was not forgotten. In 1978 the Department of Energy (DOE) funded an investigation of plants whose sugars and seed oils might be used as an alternative or supplemental transportation fuel. The DOE directed a small portion of the funds to study algae in what became the Aquatic Species Program (ASP).

ASP-funded biologists collected thousands of algae species and assessed their various growth rates, light requirements, need for nutrients, tolerance for salinity, oil production, and other characteristics. Three hundred species were deemed especially promising and went into testing in California, Hawaii, and New Mexico in labs and in raceway ponds of varying sizes.

DOE funding for the ASP averaged about $1 million per year, only about 5 percent of the funds devoted to plant-based alternatives and a paltry sum given how little we knew about algae as a possible fuel. Nonetheless, in 1995, in the face of rapidly falling oil prices and under budgetary pressure, the department closed the program and focused its remaining research dollars on producing ethanol from the sugars in corn and other plants. The ASP's final report did, however, highlight algae's economic potential, noting that "this report should not be seen as an ending, but a beginning." The report estimated that 500,000 acres could produce the energy equivalent of 8 billion gallons of gasoline, or about 16,000 gallons per acre. But the report also noted that the initial capital costs of building raceways would be high. Adding the necessary carbon dioxide to pond water to prompt that growth, separating the algae from the water, and then extracting the oil would also add to operating expenses. The bottom line: To grow, extract, convert, and then refine algae oil into usable products might put algae fuel at $240 per gallon. Of course, that was an impossible price, but everyone acknowledged that the technology was less than embryonic—you could say it was barely germinal. With continuing research and development, the price would certainly fall drastically.

Then, in the first decade of the twenty-first century, oil prices began a steady and spectacular rise, going from $17 per barrel in 1998 to $160 per barrel in 2008. Experts postulated that the world had arrived at "peak oil," the moment when global oil extraction reaches its maximum and then declines. In addition, the Kyoto Protocol, an international agreement adopted by 190 countries in 1997 to fight global warming by reducing greenhouse gas emissions, went into effect in 2005. If governments were to meet the commitments they'd made under the agreement, they would have to implement policies—carbon taxes, for example—that would make fossil fuels more expensive. Many energy experts concluded that oil prices would continue to increase indefinitely or at least remain over $100 barrel.

That assessment spurred a number of entrepreneurs and investors to take a serious—and by 2008 a downright avid—interest in algae oil. Two dozen startup companies went looking for investment funds, and venture capital firms, angel investors, foundations, and governments opened their wallets. Some of the new companies made successful stock offerings. Even major oil companies like Chevron, Shell, and Exxon stepped up.* Between 2000 and 2010, algae oil companies attracted more than $2 billion of investment.

In retrospect, it is stunning that investors ponied up so much capital on so little evidence of a profitable outcome. Although the ASP produced ideas and some interesting preliminary data, the fact is that the oil produced under the program was tremendously expensive, only eight out of thousands of algae species had been investigated in any depth, and only one method of production tried. The largest experimental ponds had been only a quarter-acre in size and there had been only two of those. Many questions of cost simply had not been addressed. Nonetheless, money rushed in.

Of all the companies in the burgeoning sector, Sapphire Energy gathered the most press coverage and funding. In early 2008, the company was working out of subleased space with just three offices, a couple of lab benches, and an unused greenhouse, but it soon secured more than $100 million in capital from Bill Gates's investment firm, Cascade Investment, as well as Venrock, Arch Venture, and the Wellcome Trust. The goal of the company was to create a "drop-in" fuel, an oil so similar to fossil crude that its refined products could be seamlessly intermixed with gasoline, diesel, and jet fuel.

---

*Some say the oil companies invested merely as a public relations matter, as a way to demonstrate environmental concern at little cost—a technique known as greenwashing. A more generous interpretation is that Big Oil was investing for the day, which seemed to be drawing near, when oil supplies dwindled.

However large its bank account, Sapphire knew that startup capital costs and salaries would quickly deplete it. The company had to move rapidly to production and a revenue stream. It leased lab, greenhouse, and administrative space in San Diego and hired staff. Its chemists looked at extraction technologies, engineers tackled raceway design, phycologists studied oil production in various species, geneticists probed chromosomes, ecologists focused on maintaining algae in the ponds, and financial types analyzed costs and returns. Sapphire—and its competitors—faced a blizzard of complex decisions. Even seemingly simple questions, like *What is the ideal velocity for paddlewheels?* or *What is the best size of carbon dioxide bubble for algae to absorb?* were difficult. The answers to those two questions, for example, were affected by each other, as well as by the cost of electricity, the shape and depth of a raceway, and, of course, the particular species under cultivation. It's not that the analyses were so exotic, it's that so many of the variables had never been studied in a lab, let alone under real-life conditions.

Sapphire's first decision was to grow microalgae outdoors in ponds rather than indoors in photobioreactors like the clear plastic panels that Valcent used. The ASP had focused on outdoor ponds, and Sapphire could build on its experience. In addition, no one had any idea what the best design for a photobioreactor might be. Researchers at universities and startups in the US and Europe were experimenting with different clear plastic structures—vertical, circular, and sloping tubes; horizontal bags; flat plates; and accordion-shaped panels—but there was no conclusive winner. Besides, at the very large scale of production that Sapphire had in mind, open ponds would certainly have a lower initial capital cost. After an intense year of lab experimentation, on December 31, 2008, researchers inoculated the first of several forty-foot-long by six-foot-wide ponds at its twenty-two-acre test facility in Las Cruces, New Mexico.

It turned out that calculating the ideal paddlewheel speed and identifying the best CO_2 delivery system was easy compared to the challenge of actually growing algae in quantity outdoors. On that front, Craig Behnke, who was vice president at Sapphire at the time, told me, "We got knocked down again and again. The most productive lab strains turn out not to be the best ones in the field. The real question, it turns out, is how does a species perform when you put it into agriculture?" Outdoors, dust storms could cloud the pond water and downpours could change its chemistry in a matter of hours. A species that was thriving under intense light in pond water with a pH of 7 would falter when the water turned murky or the pH fell. Even when a crop did finally flourish, it all too easily became a victim of its own success. Rotifers and other microscopic aquatic animals sucked up fat algae like fleets of tiny vacuums suctioning up dust particles. Algal spores from miles away floated in on the air; some that landed in the nutrient-rich ponds reproduced and ate Sapphire's selected species out of house and home. Failure followed failure. Cultivated algae succumbed to the thousand natural shocks that all crops are heir to. It was enough to drive Behnke and his colleagues mad.

The scientists and engineers had to make a career adjustment. "We had to learn how to become farmers," Behnke said. They learned, for example, that feeding algae is not like feeding plants. The life cycle of a plant takes months, so terrestrial farmers generally fertilize their fields just a few times during a growing season. An individual alga's life, on the other hand, is three days; after that, it disappears in the act of division. The algae had to be fed constantly, but in amounts (determined by trial and error) that they could assimilate quickly. Any excess fertilizer served only to attract competitors. The newly minted farmers learned how to manipulate pH and treat infestations of fungi, parasites, and hungry zooplankton. In some ways, farming algae was more challenging than cultivating vegetable or grain crops; the speed of the growth cycle is unlike any terrestrial plant. As Bryn Davis, operations manager at the time, said, "If we see something wrong in the pond in the morning, if we don't do something by lunch, it's going to be dead by the end of the day."

Testing continued through 2009 and into 2010 at Las Cruces, both in the initial small raceways as well as several new one-hundred- and three-hundred-foot-long ones. The scientist-farmers inched up the learning curve, and after more than a year of trials, they began to get it right. "Finally, we got to the point where we could put in a new species and grow it to large scale pretty quickly," Christopher Yohn, associate director, told me. It was time to move to a commercial scale.

In 2009, the Department of Energy awarded Sapphire $50 million and the company obtained a $54.5 million loan guaranty from the Department of Agriculture to build a hundred-acre commercial facility in Columbus, New Mexico, where land was cheap and sunlight abundant. The Columbus operation would not only grow algae but also convert it into crude oil. (The crude was then sold to traditional refiners who cracked it, as they do fossil crude, into the various kinds of fuels that different vehicles require.) In 2012, Sapphire put three algae species into mass production.

All this time, engineers had been focusing on how to get the algae out of the water and then the oil out of the algae with the least energy expense. That question was, and still is, critical: 40 percent or more of the energy cost of producing algae oil lies in two operations, harvesting and processing. In outdoor ponds, growers hold the algae density to only about half a teaspoon of algae per gallon of water. Otherwise, the algae would shade each other and undermine the growth of the overall population. But this low density poses problems. While some species float at the water's surface or naturally sink to the bottom, many others have evolved to be weightless, rising and falling slowly in the water column. So, what's the cheapest way to get algae out of water?

Mechanical filtering is an obvious choice, but filters fine enough to trap microalgae, which are narrower than a hair's width, are all too easily clogged and need frequent cleaning. Alternatively, you can aggregate the algae first – it's easier, after all, to trap one dust bunny than millions of dust particles. Some growers add microscopic particles to the algae broth in a process called flocculation. The algae adhere to the particles, forming clusters or "flocs" that either float to the surface or sink, depending on the density of the particular particle used. But then you have to separate the flocculant from the algae. Or, you can spin the water out of algae with high-speed centrifuges. Growers choose a method by assessing a number of characteristics, including whether the algae are floaters or sinkers, and sometimes use different methods in combination.

Once the algae has been dewatered, it's time to extract the oil from the cells. This, I assumed, would be the easy part; after all, the cell walls of microalgae are vanishingly thin. How hard could it be to breach them? Very hard: To survive ever-changing levels of hydrostatic pressure as they drift up and down in the water column, microalgae have evolved cell walls that are remarkably tough yet elastic — and highly resistant to rupture. And that's not the only obstacle to reaping their oil. Dewatered algae aren't dry algae; the biomass still contains a lot of water, which acts as a lubricant. If you imagine trying to smash a bunch of bath oil beads in a bowl of water, you'll see the problem: most of the beads will squirt away from under your hand, intact.

You could heat the algae to dry them and then use a standard oil press, but dryers require a lot of energy. More often, growers treat the algae biomass with a solvent like hexane, which chemically dissolves the cell walls and binds to the oil. That's effective, but then the oil has to be boiled to separate it from the hexane, which adds still more expense. And hexane is a hazardous chemical.

Other, more experimental methods include popping the cell with sonic or microwave energy or adding chemicals that lyse (break apart) the cells and release their contents. But whatever method growers use, getting the oil out of recalcitrant algae is a major driver in the cost of algae fuel.

Given the expense of extraction, would it be possible to avoid it altogether and simply turn the whole algae into oil? After all, nature made petroleum from whole algae with pressure and the passage of millions of years. In 2009, Sapphire's engineers began testing a method that turns all of the organic parts of the algae into oil. Instead of taking eons, they could do it in hours.

The method is called hydrothermal liquefaction, or HTL. In HTL, biomass is treated with high-temperature water (about 660 degrees Fahrenheit or about 349 degrees Celsius) in a chemical reactor pressurized to about three thousand pounds per square inch, about two hundred times the pressure at sea level. The pressure prevents the water from vaporizing; instead, it becomes "superheated." Superheated water is corrosive, and it rips apart solid organic compounds. HTL is an appealing method for processing algae because it works on wet biomass, thus mooting the expense of drying. "HTL is like a molecular blender," Behnke says. "Nothing even vaguely looking like a large, intact biomolecule makes it through the process." Because all of the algae's organic components are liquefied, not just the oil, HTL had the potential to double the company's crude oil production. The crude did still contain phosphorus and other minerals that had to be removed (at an additional expense), but they could be recycled as feed for more algae.

HTL has been around since the 1920s, but until recently no one had built a system capable of working at a meaningful scale or over a long period of time. Test systems were plagued with mechanical problems, and the metal parts in the HTL machinery failed in the corrosive environment. Operators had to add biomass gradually, in small, discrete batches, and stop to clean the processor frequently. And HTL was dangerous: imagine the scalding mayhem caused by a boiler explosion, and multiply that a thousandfold. Nonetheless, Sapphire developed a patentable, continuous process for HTL in 2011 and installed a small pilot unit at its Las Cruces site.

That year, the *Wall Street Journal* highlighted Sapphire for its potential to produce "green crude," and *Forbes* picked the company as one of sixteen early-stage "companies to watch." The Environmental Protection Agency certified that the company's crude could be put through traditional refineries and met all the requirements under the Clean Air Act. In 2012, the company had the largest raceways ever built (at 2.5 acres each) and became the first outdoor algae farm in the world to deliver a significant quantity of oil. In the fall of that year, the US Navy, as part of its effort to diversify its fuel sources, powered the Great Green Fleet, a strike force comprising dozens of jets and helicopters, two destroyers, and a cruiser, with algae oil and used cooking oil mixed in a fifty-fifty blend with petroleum fuel. In 2013, Sapphire reported it had generated one million gallons of crude oil thus far. Then, Tesoro Refining and Marketing Company (now Andeavor), a leading independent oil refiner, signed an agreement to buy its crude. Tim Zenk,

Sapphire's corporate affairs director, announced at an industry conference in September of that year, "The algae fuel industry is moving steadily on its path towards commercialization... Policy makers can confidently project the sustainability of fuels produced from algae."

Sapphire was not the only company providing algae oil for the Great Green Fleet. Solazyme, a San Francisco–based public company, was too. In the summer of 2016, I head to California to meet with Jonathan Wolfson, cofounder of Solazyme — which had been renamed TerraVia shortly before we spoke — in a conference room at its headquarters. Wolfson, a brown-haired, blue-eyed, stocky man in his mid-forties, is in a hurry, talking emphatically, setting up a visit to the company's test kitchen for me, trying to squeeze in our interview before a meeting, and glancing at his phone, waiting for a call from his wife telling him to drive to the hospital for the birth of their third child.

He recounts Solazyme's origin story. He and Harrison Dillon, a buddy at Emory University, both loved the outdoors. Their hiking adventures cemented their interest in environmental issues, and they dreamed of doing something meaningful to protect what they loved. As undergraduates in the early 1990s, they kicked around various ideas for eco-friendly businesses, but after graduation they went their separate ways. Dillon earned a PhD in genetics and a law degree from Duke, and he worked for a firm in Silicon Valley for a few years. Wolfson got an MBA and a law degree from New York University and started a financial services software company. But the friends reunited in 2003, at a time when both had the education and business experience to do something, as Wolfson puts it, "that really matters and you can feel good about and at the same time make money." They decided that the thing they would do was algae fuel. At the time, the Department of Energy was still funding scientists through the Aquatic Species Program, but no one had yet tried to make a go of a company. Wolfson moved to Silicon Valley and the two started Solazyme in, of course, a garage.

At first, like the DOE-funded scientists, they focused on algae that would grow in outdoor raceway ponds using solar energy. With the backing of family, friends, and angel investors, they hired six researchers to test the oil output of various species and to tinker with their genomes to see if they could be coaxed to make more, and more crude-like, oil. For the first few years, they thought they were making good progress on the scientific front. But one day, Wolfson said, they took a hard look at the numbers and realized that, while they would be able to produce oil, they wouldn't be able to sell it anywhere close to the market price of fossil fuels. With considerable trepidation, they decided to change their fledgling strategy and went to their investors with a completely different approach.

"What we figured out," Wolfson said, "is that photosynthetic algae are not, on a per-acre basis, terribly efficient. They depend on access to light, which means they need a *lot* of area to spread out and not be shaded by each other. On the other hand, what they *are* really efficient at is the dark side of photosynthesis." What Wolfson is referring to is the light-independent half of photosynthesis, known as the Calvin Cycle. After algae make ATP with the power of sunlight, they use that temporary store of chemical energy – regardless of whether it is day or night – to transform carbon dioxide into simple sugars, and then into all kinds of more complex organic compounds, including oils. "We didn't see that at the beginning, but once we did, we started looking at species that *don't* photosynthesize but are great at making oils."

Wait – algae that don't photosynthesize? Isn't photosynthesis the very essence of what it means to be an alga?

Recall that the mother of all algae was a single-celled heterotrophic organism with a flexible membrane. That organism engulfed a photosynthetic cyanobacterium but didn't manage to digest it. The cyanobacterium's descendants lived on inside the heterotroph's descendants as chloroplasts, which power their hosts with energy from the sun. This fantastic innovation led to the greening of the planet and the creation of us.

But, eating the sun isn't always the best solution for all microalgae. Some species retained – and have reactivated – their ability to eat bits of organic matter for energy, just as their heterotrophic forebears did. Like hybrid cars that can run on gas or electricity, these mixotrophs choose their fuel, depending on whether photons or organic molecules are most abundant.* Over the long course of Earth's history, a handful of microalgae species found themselves in circumstances where eating organic molecules was always more efficient than absorbing sunlight. These species — no more than fifty have been discovered so far — abandoned the solar experiment altogether, deactivated or eliminated their chloroplasts, and sent themselves back in time to become heterotrophs once again.

Wolfson and Dillon focused on finding the most efficient of these heterotrophic algae species. They were pioneers in the area; few others had investigated these algae for their oil-producing potential. Not only did they find good candidates that naturally could store about 50 percent of their volume in oil, they developed a bioengineering toolkit for tweaking their genomes to improve that percentage, up to as much as 80 percent. Then, using technology already highly refined in the beer, ethanol, and other fermentation industries, they set about growing their algae in closed steel vats, feeding them sugar directly.

Wait — I interrupted Wolfson — isn't the great thing about algae that they use free, non-polluting solar power for their energy? Why feed algae sugar that Solazyme has to pay for? Besides, plants are much less efficient than algae at turning photons, carbon dioxide, and water into sugars. And in order to grow, harvest, and process a terrestrial sugar crop — whether it's corn, sugar beets, or sugarcane — you have to use arable land, scarce fresh water, and costly fertilizers, as well as energy to seed, till, harvest, and process the plants. How could the oil of sugar-fed algae compete in the marketplace with, say, Sapphire's oil made from algae that use free sunlight? And how could the method be environmentally sound?

---

*Why aren't all algae mixotrophs able to switch-hit? The flexibility has a cost: mixotrophs have to invest in building and repairing the machinery for both energy systems — chloroplasts and also digestive organs and enzymes. Nonetheless, mixotrophy is an effective strategy; recent research reveals that there are far more mixotrophic species than previously thought.

Wolfson had answers. Algae eating sugar and living in steel tanks can grow around the clock, not just during daylight hours. They can thrive in any climate in the world, not just in warm, sunny regions. Grown in stories-high fermenters, they produce vastly more oil per acre than algae grown outdoors, and they use far less water. In addition, he has few of the issues that outdoor growers have because the cultivation in tanks is completely controlled. No competitor species drift in; no hungry zooplankton chow down on the crop; and no sandstorms, cool mornings, cloudy days, or rainstorms slow production.

Moreover, Solazyme's algae grow ten times more densely in fermenters than they do in outdoor ponds. That means Solazyme saves money in processing its crop. The algae soup is simply dried and sent through ordinary oil presses, the same low-tech, inexpensive machines that canola, soy, and other seed-oil processors use. "This significantly reduces the amount of electricity we use," Wolfson said. In the final analysis, although buying sugar to feed algae is an additional operating cost, Solazyme offsets that expense with higher productivity and reduced energy expenses.

In December 2007, Solazyme produced its first renewable fuel oil; by the end of the decade, it had ramped up production in its own small fermenter facility in Peoria, Illinois, and in contract fermenters. In 2010, the company made its first sale – twenty-two thousand gallons of fuel – to the US Navy; in 2011, its fuel partly powered a United Airlines jet; and by 2012, it was providing algae fuel to the Great Green Fleet, just like Sapphire. The company was, as the Solazyme mantra had it, "taking the world's smallest organisms to solve some of the world's biggest challenges."

But then, in 2014, disaster struck for Sapphire, Solazyme, and the many other smaller companies in the algae fuel business. Hydraulic fracking, a new oilfield drilling technique, ramped up American crude oil production, and foreign producers were unwilling to reduce theirs. As a result, crude oil prices plummeted. The price per barrel, which had climbed to $120 in 2011, fell to $70 in 2014 and reached $30 at the end of 2015. In many places around the US, gasoline was selling for under $2 per gallon. Algae fuel couldn't compete.

As Sapphire's Yohn put it, "We had been prepared to compete with oil at $80 or $90 a barrel. But it's impossible for *any* alternative fuel to compete at $30 a barrel." In 2015, Sapphire closed down its algae fuel operations. From a peak of 140 employees, Sapphire had retained about 40 in February of 2016 when I visited. Like many former algae fuel companies, the company had by then recognized that its survival depended on producing higher-value, lower-volume products. When I spoke with Behnke and Yohn, they were trying to pivot to pigments, proteins, nutraceuticals, and fish and animal feed. Indeed, a number of the smaller algae oil companies have managed to survive in this manner; several are now producing astaxanthin, which is a carotenoid pigment added to animal feed to color egg yolks and salmon flesh, and as a human food supplement. Ultimately, however, Sapphire was unsuccessful and closed down entirely in 2017.

Solazyme, Wolfson told me, ultimately produced well over a million gallons of algae fuel oil. "With tax incentives, we were breaking even" – not bad for a startup company with a previously untried technology in a new industry. "But right when the business started getting interesting, oil prices collapsed." Solazyme was still selling algae fuel to the package delivery company UPS in 2016, but it had to put its grand vision of replacing fossil fuels with renewable algae fuel on hold. Fortunately, it had already turned its algae technology in new directions.

6

The Algae's Not for Burning

As soon as Solazyme's bioengineers began tweaking algae's genomes, they realized that their tools allowed them not only to produce high-quality oil to burn in vehicles, but *any* oil that manufacturers or consumers might want. By tinkering with the genes that control the length and shape of its hydrocarbon molecules, they could alter the properties — say, the viscosity or degree of unsaturation — of any algal oil. Algae could produce cosmetic, lubricating, cleansing, or edible oils.

From the outset, Solazyme saw the financial potential of "designer" oils. For one, they are not undifferentiated commodities like crude oil, which means their prices are not dictated by a global market. Even better, their prices per ounce are orders of magnitude higher than fuel oil. And, while growing genetically engineered algae outdoors could raise environmental concerns, modified algae grown in closed fermenters posed no risk at all.

In 2011, well before the crash in oil prices, Solazyme had already signed a joint development agreement with Dow Chemical Company to provide oils for use in insulating fluids for electrical transformers. Solazyme's dielectric oil had a particularly high "flash point" — the temperature at which its vapors ignite — a great benefit when you're playing with high-voltage electricity. That same year, the company introduced Algenist, a skin-care product containing a compound Solazyme calls "alguronic acid." Alguronic acid, the company claims, is a mix of polysaccharides that have helped algae protect themselves from environmental stresses for billions of years and has an antiaging effect on human skin. (Keep in mind that cosmetic companies needn't provide scientific evidence for such claims.) A single ounce of Algenist sells today for

$115; compare that to an ounce of crude oil, which sells for pennies. Algenist, sold by Sephora and QVC, has been a financial winner for the company. Solazyme recently sold the brand to Tengram Capital Partners for about $20 million. In 2016, Unilever signed a $200 million, five-year purchase agreement for the company's tailored oils to incorporate into its personal-care products.

Solazyme also recognized that its heterotrophic microalgae can produce the long-chain omega-3s (DHA, as well as eicosapentaenoic acid or EPA) that are important for human health. As we've seen, infants need DHA to develop full brain capacity, and they consume it in breast milk and baby formula, which is universally fortified. And adults, who need omega-3s for both cardiovascular and brain health, get theirs from eating coldwater fish like tuna, halibut, cod, and salmon, which accumulated the oil by eating smaller fish that subsisted on algae.

But wild coldwater fish have become increasingly expensive for consumers, a function in large part of diminishing supply as well as rising demand as people have come to understand the health benefits of eating them. Fifty years ago, the oceans were full of these large fish; no one imagined they could be depleted. But, with the advent of modern trawlers using sophisticated fish-detection equipment and dragnets a half-mile long, we humans have turned oceans into ponds and ransacked them of their larger denizens. The UN Food and Agriculture Organization (FAO) reported that the share of fish stocks that are being taken at unsustainable levels rose from 10 percent in 1974 to 31 percent in 2016. Our species is pushing the limits of the oceans. The global wild fish catch has been declining, from a peak in 1995 of 130 million tons to 110 million tons in 2010 and heading toward 90 million tons in 2020. You can see the results in higher prices at the grocery store.

Despite this trend, global fish consumption doubled between 1960 and 2013, thanks (many thanks!) to fish farming. In 2014, for the first time, we bought more farmed fish grown in aquaculture — chiefly salmon, tilapia, carp, and catfish — than those plucked from the wild. This is a generally positive development for people and the planet. For one, without farmed fish there would be even greater pressure on wild stocks, and higher prices would put fish — and its protein and nutrients — out of reach for many consumers.

Still, farmed fish need to be fed, and fishmeal and fish oil made from forage fish are the best foods for them. But forage fish are being ever more intensely fished and sometimes overfished. When there are too few of these prey fish, sea lions starve, seabird populations fall, and predator fish become scarce. It's true that forage fish populations can naturally fluctuate from year to year, which can make it hard to be certain whether a scarcity is due to fishermen or Mother Nature. Nonetheless, according to a report published by the National Academy of Sciences in 2014, when forage fish stocks are *already* at a natural low point, overfishing creates more frequent and more dramatic declines. In recent years, the Peruvian government has had to close its offshore anchovy and sardine fisheries because the fish populations were so depleted they might not have recovered from further harvesting.

While forage fish reproduce quickly and most populations will bounce back, the effects on the larger fish and seabirds that eat them are not clear. The top-tier predators mature and reproduce more slowly; their populations may not rebound if they suffer years of diminished prey, especially if they are already overfished. There's another reason to protect forage fish: the Lenfest Forage Fish Task Force, funded by the Pew Charitable Trusts, reported in 2012 that forage fish are worth twice as much if they're left in the ocean as food for more valuable table fish than if they're removed and turned into feed.

The production of farmed fish is increasing at a rate of nearly 10 percent per year. At the same time, forage fish are becoming scarcer and their prices are rising, which means, as we've seen, aquaculturists are feeding more plants to their stock. That creates two problems: more poorly digested plant food ends up in the water, where it causes environmental problems, and the level of omega-3s in the fish we eat declines.

There is a way to solve — or at least alleviate — these problems. Since DHA in wild carnivorous fish comes from the algae they consumed by eating forage fish, why not simply feed farmed fish directly with algae? In other words, why not cut out the middlefish?

Entrepreneurs, including Solazyme, are working to do just that. In 2012, Solazyme entered into a joint venture in Brazil with Bunge Limited, a global agribusiness and food company headquartered in New York. Bunge contributed its sugarcane plantation and mill located about 330 miles northwest of São Paulo; Solazyme contributed its technology. The joint venture, Solazyme Bunge Renewable Oils (SBO), owns and operates a new $200 million fermenting facility that uses sugar from the adjacent mill to feed heterotrophic algae. The facility opened in 2014 and can manufacture up to one hundred thousand metric tons per year of algae oils and powders. So far, DHA has been its chief product.

There's no other venture quite like SBO, so I'm pleased when David Brinkmann, vice president of engineering technology, and Jill Kaufmann Johnson, vice president for sustainability and external affairs, offer to take me on a tour, via drone camera, of the company's facilities. As I fly in from the south, I see new white-roofed buildings nestled against the rusty, corrugated tin roofs of the old sugar mill complex. Beyond the complex are miles of dense green sugarcane fields interspersed with groves of eucalyptus and rubber trees, cornfields, and grazing land, as well as stands of trees native to the savanna-like climate of this part of Brazil. We zoom in toward the heart of the SBO operation: six, eighty-foot-tall steel silo-style fermenters veiled in a shining tracery of steel pipes and girders. On top of each fermenter is an aquamarine engine housing. Bright yellow access staircases and catwalks zigzag up and across the fermenters.

The SBO camera — now in the hands of a human operator — ushers us into a laboratory building where a technician wearing heavy gray gloves extracts a thumb-size vial of algae from a cryogenic freezer and transfers the contents to a tabletop fermenter. The microalgae in the vial are a species of *Schizochytrium* originally discovered in a mangrove swamp in Florida. In a broth of sugar syrup, nitrogen, and a host of micronutrients, and under optimal pH and temperature conditions, the cream-colored algae multiply rapidly, and the pale soup is then transferred into ever-larger vats. When the algae are numerous enough, they are put in the main fermenters, where they continue to multiply. The liquid inside the fermenter is continuously mixed and exposed to cooling surfaces; otherwise, flush with food and oxygen and busily growing and reproducing, the algae generate so much heat they'd cook themselves.

When the algae reach the right density, the nitrogen input is cut off. Without nitrogen, they stop growing and dividing and instead go into self-preservation mode, converting sugars into energy-dense oils — they're hunkering down for hard times. If more propitious conditions were to return, they would transform the oil back into more readily metabolized sugars, but they'll never get the chance. After several days of nitrogen deprivation, when the algae have become bloated with oil, it's harvest time. The broth is piped from the fermenters to standard industrial dryers that evaporate the water, leaving behind a dry, golden or beige whole-cell algae powder. The powder is packaged in bags and barrels, then shipped to aquaculture and animal-feed manufacturers who combine it with other ingredients and pelletize it into feed.

The algae are like Rumpelstiltskin's spinning wheel, transforming Bunge's cheap sugar into high-value omega-3 oil. When it's fed to farmed fish, SBO's oil spares the lives of wild forage fish while maintaining the DHA content of farmed fish. And if more algae-based feed oil means less corn and soy in the fishes' diet, then the environment benefits, too.

But is the happy ending deceptive? Sugarcane plantations in Latin America and East Asia have a bad reputation. All too often, they are carved out of tropical forests and pollute freshwater ecosystems with eroding soil and excess fertilizers. Milling operations are known to dump residue into local waters, and the fields themselves are dangerous places. The laborers who work in them — including, in some countries, children — contend with machetes and sharp stalks that pierce their skin, long hours, low pay, and an increased risk of kidney disease from becoming dehydrated while working in the intense heat. So, it's fair to ask: what are the environmental and human costs of these sugar-fed algae products?

Several years ago, the World Wildlife Fund established a rigorous certification program called Bonsucro that sets standards for environmental and worker protection at sugarcane fields and mills. For many years, Bunge's Moema sugarcane mill and many of the surrounding plantations have secured Bonsucro certification. The plantations are more than a thousand miles from the rainforest, and the whole area has been farmland for decades. The complex uses no fossil fuels for production: sugarcane stalks and leaves are fed to boilers that provide the energy. Sugarcane is a thirsty plant and has the potential to draw down underground aquifers, but these plantations are entirely rain fed, and the excess water from the fermentation tanks is recycled onto the crops. The bottom line: DHA made from the Moema mill's sugarcane has low environmental impact. In 2016, BioMar, a Danish feed company, signed an agreement to buy SBO's omega-3 products, and the industry press reports some forty thousand metric tons have been delivered so far.

It's not just fish that eat Solazyme's edible oils. Wolfson, Dillon, and others recognized that microalgae have potential as healthy replacements for oil and proteins in human foods, too. Early on, an employee decided, on a lark, to make a cake with some of the company's *Chlorella* algae, figuring he might be able to substitute the algae for the butter and eggs in his recipe. The cake, served at the weekly all-company meeting, was a toothsome success, and it led to further employee experiments (brownies and banana rum cake are fondly recalled).

Solazyme—which I will henceforth call TerraVia because its name was changed as it expanded its food focus—developed two whole-algae products: a high-protein powder and a high-oil powder that is low in saturated fat and cholesterol. Both are made with non-bioengineered algae and are vegan-friendly. The company now sells the golden-hued powders to food manufacturers, who incorporate them into protein bars, brownie mixes, egg substitutes, salad dressings, and other packaged foods.*

---

*When I visited TerraVia, the powders were made with algae fed with corn-derived sugar at its facility in Peoria, but the company is considering alternative sugar sources, including sugarcane.

A new source of protein and healthy cooking oil sounds good, but what about the taste? Wolfson takes me to the company's test kitchen. There are a number of dishes on a counter, presented in pairs, one made with TerraVia powders and the other with standard ingredients. But before I get to the crackers, cookies, and French fries, I'm invited to try a little of the algae powders straight from ramekins. I'm a little queasy at the thought, but I take a breath and try a teaspoon of the protein powder first. The powder is hyperfine, and I roll it around in my mouth, ready to grimace. But no, the taste is almost neutral, with a slight hint, to me, of peanut, and the same mouthfeel as rich chocolate. The oil powder is equally fine in texture and has just the slightest bite to it. I'm greatly relieved; there's no fishy taste at all.

Next up are slices of challah bread, one made with vegetable oil and eggs and the other with oil-rich algae powder, no eggs, and only a little vegetable oil. I find no difference in taste or texture, but the bread made with the algae powder has no cholesterol, one-third the fat, and 20 percent fewer calories. The algal chocolate chip cookies are as tasty as their conventional counterparts. A mayonnaise made with an oil-rich whole algae strikes me as having a more neutral taste than the eggy original. I try cheese crackers made with algae protein, which doubles their protein content, and they taste the same as the crackers made without algae.

The algae used in edible powders are not bioengineered, but TerraVia does alter the genomes of *Prototheca* species to produce a nearly infinite variety of specialized edible oils. If you have qualms about GMOs, there's no reason to worry: Only the algae genome is modified — the extracted oil has nothing bioengineered in it. (In much the same way, bioengineered bacteria make human insulin and the clotting factor for many cheeses.) One such oil is TerraVia's algae cooking oil. Marketed as Thrive, it is colorless and tasteless, which makes it good for cooking pancakes, frying French fries, or making any other dish that doesn't feature the flavor of the oil. It also has the highest percentage of unsaturated fats of any cooking oil, as well as a high smoke point, which is helpful on the stovetop. The Bon Appétit restaurant chain uses Thrive in more than 650 of its cafés, and Walmart now stocks it.

• • •

For the most part, Wolfson tells me, when it comes to its bioengineered oils, TerraVia responds to whatever food or personal-care companies require, from palm oil alternatives to cosmetic oils. The possibilities are infinite. The company just introduced a "structural fat," which they call algae butter, as a healthier—and vegan—replacement for semisolid, highly saturated fats like butter and palm oil used in baked goods.

Despite the company's ingenuity, however, its road to financial success has been rocky, to say the least. A joint venture with a French company, Roquette Frères, ended with Roquette's theft of TerraVia's intellectual property, followed by successful but expensive litigation. In 2017, overburdened with debt, TerraVia went into chapter 11 bankruptcy and was purchased by Corbion, N.V., a Netherlands biotechnology company with annual revenues of more than $1 billion. With its focus on food and bio-based ingredients, Corbion appears to be a good home for TerraVia. The fact that sales of vegan-friendly foods are on the rise and consumers are trying to cook more healthfully bodes well for success.

Corbion is not the only player in the market for algae-derived omega-3s. DSM, another Dutch company, has long used heterotrophic algae to produce DHA that is added to infant formula and other foods. Hawaii-based Cellana and Qualitas Health in Texas grow algae in outdoor ponds for omega-3s that are sold primarily as vegan substitutes for fish oil supplements.*

Still, companies that produce algae for fish feed are attracting the largest investments. Skretting, the world's largest producer of

---

*While the evidence is clear that eating oily fish species is good for cardiovascular health, a meta-analysis published in *Cochrane Database of Systematic Reviews* in July 2018 that involved more than 112,000 people concluded that taking long-chain omega-3 oil supplements probably makes little or no difference to the risk of cardiovascular events, coronary heart disease events or death, or stroke. And, as Paul Greenberg, author of *The Omega Principle: Seafood and the Quest for a Long Life and a Healthier Planet,* explained to Terry Gross, host of the National Public Radio program *Fresh Air,* as much as ten million metric tons of Peruvian anchovies—about one-eighth of all the fish caught in the world each year—goes into omega-3 oil supplements. Even if you choose to take the supplements, an algae-based version would be a better choice for the oceans' health.

aquaculture feed, is creating its own algae-based feed. A partnership between Arizona-based Heliae and Washington-based Syndel is producing another, called Nymega. Evonik, a German company, and DSM have started a joint venture, Veramaris, that will produce omega-3s from marine algae in a $200 million plant in Blair, Nebraska. As Michael Tlusty, the director of ocean sustainability at the New England Aquarium, has said, "In coming years, we're going to need forty million metric tons of seafood annually to feed two billion additional people." We're going to need algae to get there.

7

Ethanol

I first encountered algae fuel in Valcent's Texas greenhouse. I was intrigued from the start by their photobioreactors — those rows of clear plastic panels with serpentine channels of algae and water running through them. Valcent's cultivation method, assembly line–like and isolated from the natural environment, seemed matched to the world's relentless demand for liquid fuel. So, despite the company's sudden demise, I followed the fortunes of two other companies, Joule Unlimited and Algenol, that were growing algae — by all accounts successfully — in their own versions of photobioreactors. When I started my research for this book and contacted the two companies, I found Joule's management unapproachable, but Paul Woods, founder and CEO of Algenol, was anything but reticent.

Which is why, in July 2013, I am standing with him under a blazing Florida sun at the company's headquarters, situated on a sandy, scrubby lot not far from the Fort Myers airport. Woods is dressed in shorts, a T-shirt, and sandals, and his strawberry blond hair, falling from a high forehead, reaches his shoulders. Not a typical CEO look, but Woods is not your typical CEO.

In front of us are row upon row, rank upon rank — altogether more than two acres — of glowing, green, plastic pouches, each four feet wide and four feet tall and only a few inches deep. Seven thousand of them hang vertically on steel frames, spaced about ten inches from one another and a foot off the ground. In total, the array looks like a field of hanging files for an office of giants. But instead of papers inside each file, there are about four gallons of sterilized saltwater, bubbles of carbon dioxide, and millions of bioengineered cyanobacteria. The transparent files are Algenol's patented version of photobioreactors, otherwise known as PBRs.

What those cyanobacteria are producing is not oil, but ethanol. Read the labels on the pumps at the gas station and you know that petroleum isn't the only liquid fuel that can power our cars. Most gasoline sold in the US is 10 percent ethanol made from fermented corn. Ethanol — otherwise known as ethyl alcohol — is the intoxicating component in your pilsner as well as the bacteria-killing compound in Purell and other disinfectants. But compress it in an engine's cylinders, combine it with oxygen, ignite it with a sparkplug, and like petroleum, it suddenly releases energy held in chemical bonds. Powering vehicles with ethanol is not a new idea; the first internal combustion engine ran on it. Today, transportation fuel in Brazil is about 25 percent ethanol made from fermented sugarcane. Almost all cars in Europe can also run on at least a 10 percent ratio of ethanol to gasoline. Flex-fuel cars in North America and Europe can run on any combination of a blend of ethanol and gasoline up to 85 percent ethanol.

Today, ethanol is produced by yeasts as they ferment corn, sugarcane, and other plants; the ethanol is simply yeast pee. Some bacteria also live by fermentation: that's how cabbage becomes sauerkraut and cucumbers turn into pickles. But fermentation has never been in cyanobacteria's life plan — at least not until Algenol convinced them to reconsider.

Woods invites me to take a closer look at the PBRs. Each plastic pouch has ports, like those on an IV bag, for adding carbon dioxide and nutrients and for removing ethanol and excess biomass. "We get these PBRs to go from green to nearly black with cyanobacteria in the space of four days," Woods says. Then, their contents are pumped to distillers and steam strippers where the ethanol is separated from water and concentrated, and the biomass is collected. Technicians clean the PBRs in place, refill them with saltwater, and inoculate them with the next batch of algae. (The inoculum is grown in vats indoors.) "The beautiful thing is nothing goes to waste. At the end, there's not only ethanol for fuel, but biomass that's full of useful proteins for animal feed, as well as nitrogen and phosphorus that can be used for fertilizer. There's fresh water, too. Saltwater goes in, but drinking water comes out, so Algenol is also a desalination plant."

We walk back to the headquarters building so we can see the laboratory—thirty thousand square feet of lab benches covered with blinking equipment attended to by scientists and technicians in white lab coats. Algenol employs about 120 people, including chemists, microbiologists, and process engineers, among others. One team explores how to improve the design of the photobioreactors and their arrangement outdoors. Others focus on new ethanol separation technologies, improving the growing conditions inside a PBR, and, above all else, bioengineering cyanobacteria.

Bioengineering is at the heart of Algenol. Algenol's microbiologists have inserted certain yeast genes into cyanobacteria's genome, which enable them to ferment the sugars they make via photosynthesis and trickle out ethanol. It's a neat bioengineering trick to pull off. Not only do the cyanobacteria do something they've never done in their more than three billion years of life, they are hardy enough to survive their own toxic waste—ethanol is as deadly to cyanobacteria as Purell is to the bacteria on your hands.

Not surprisingly, it has taken years to develop the Algenol process. Woods started down this road while a genetics major at the University of Western Ontario in 1984. When I asked him about the birth of his idea, he said—and this is so like Woods—"It's May, and my lab partner in a genetics course says to me, just out of the blue, 'You know, ethanol is the future of fuel.' Of course, we had both just lived through the oil embargo and lines at the gas pumps. *OK,* I thought, *that makes sense.* Then three weeks later I was sitting at my window, looking out at the green lawns and trees, and I had the stupidest idea: What if our skin could make chlorophyll and we could make all our own food, our own sugars, from the sun? We wouldn't have to waste so much time eating. Of course, that's completely impractical because we don't have nearly enough skin surface. But that made me think about algae and how they're completely green, completely photosynthesizing, and how nothing on the planet can generate sugars faster than algae. And suddenly, I thought, *If we could just get algae to use those sugars to make etha-*

nol. And it was literally this tiny, infinitesimally short thought, that you should use algae to make ethanol, that got all this started."

But, of course, algae don't make ethanol, and he knew they would have to be genetically engineered to do so. He also guessed that cyanobacteria would be the best kind of algae for the job because, like all prokaryotes, their simple circular chromosomes have a natural tendency to incorporate other bacterial genes, which makes them easier to manipulate than eukaryotes.

After college, he put together a business plan for making ethanol from cyanobacteria and tried to sell Canadian fuel producers on the idea. The companies were decidedly uninterested. For one, while the world market today for ethanol is twenty-two billion gallons a year, in 1984 it was a mere nine million, hardly any of which was used as fuel. And genetic engineering was in its infancy then; no one had yet tried, much less succeeded, to bioengineer a cyanobacterium. Nonetheless, despite having neither a proven technology nor a market, Woods raised $200,000 from friends and acquaintances for a company he named Enol Energy.

With money in hand, he set out to find someone who could bioengineer a cyanobacterium. He set his sights on Dr. John Coleman at the University of Toronto, one of the few Canadian scientists who had any experience at the time in genetic engineering. Coleman told me that Woods first phoned him out of the blue at his laboratory. "I was working on an antimalarial project at the time, trying to add genes to a cyanobacterium that would reduce mosquito populations. Mosquito larvae eat cyanobacteria, and I was trying to get cyanobacteria to express proteins that would kill or sterilize any larvae that ingested them. I thought it was quite possible we could do what he wanted."

Maybe so, but it wouldn't be easy. Gene splicing in the early 1990s was tedious, scattershot, and unreliable. No one had even fully sequenced the genome of a bacterium, and wouldn't until 1995. For years, Coleman, working with another University of Toronto scientist, Dr. Ming-De Deng, tried to insert two "packages" of yeast genes into effective places in a cyanobacterium's chromosomes. (The insertion points are critical: putting a gene in the wrong place in an organism's genome can disrupt one of its critical functions, killing the organism.) After each attempt, they would test to see if the algae excreted any ethanol. For years, nothing worked. It was as if Coleman and Deng were bicycle messengers, trying to deliver packages to an unknown address in a large city, at night and without a map, just by asking people along the way, "Are we there yet?" But in 1996, after eight years, 2,100 trials, and some good luck, they finally got the yeast genes into the right places.

I asked if they'd had a big celebration. "No," Coleman replied, laughing. "Nothing had burst into flames. The engineered cyanobacteria expressed just a tiny amount of ethanol. Still, it was a bit of a eureka moment." In 1997, Coleman and Deng published a paper on their work, and Woods applied for the first patents.

In the meantime, Woods had made a fortune as an entrepreneur in brokering natural gas and he'd retired to Florida at the age of thirty-eight. There, he told me, he spent five years contentedly sitting on the sidelines, driving a white Rolls-Royce, and traveling. But by late 2005, the price of oil was rising. Better yet, the market for ethanol had been transformed by the 1995 Renewable Fuel Standard (RFS) legislation, which required, in brief, that gasoline contain a certain volume of renewable fuel, which as a practical matter meant that 10 percent of gasoline had to be ethanol.

Legislators originally mandated the RFS because ethanol additives reduce dangerous carbon monoxide levels in tailpipe emissions. They also hoped that by substituting ethanol for 10 percent of the petroleum in gasoline and diesel fuels, they could reduce American dependence on imported oil. But in the first decade of the twenty-first century, the RFS took on a new role, as a tool in the campaign to limit human-generated carbon dioxide emissions that are warming the planet. While vehicles burning ethanol still emit CO_2, that CO_2 was taken out of the air by photosynthesizing organisms as they grew. Burning ethanol in fuel recycles atmospheric carbon dioxide. The more ethanol there is in gasoline, the less new carbon dioxide from long-buried petroleum enters the atmosphere.

The easiest way to make large amounts of ethanol is by fermenting corn, and by 2005, the US was producing about six billion gallons of corn ethanol per year. But environmentalists were beginning to point out that transforming corn into ethanol involves transporting a bulky commodity from farms to fermentation plants, and then milling, heating, liquefying, distilling, centrifuging, and denaturing it – all processes that involve burning fuel. Growing corn also requires fertilizers made from natural gas. People began to question how much we were really reducing carbon dioxide emissions by using corn ethanol. In addition, corn is a notably thirsty plant, and the crop is a drain on the shrinking Ogallala aquifer that provides most of the water available for farming in the Midwest. Some also argued that the RFS diverts corn to biofuel and away from edible products, driving up the cost of food and feed.* All of which is to say, interest was growing in making ethanol from other biomass, be it cornstalks, switchgrass, wood chips, or something else.

That something else, Woods hoped, would be algae. The time had come, he decided, for a new run at algae ethanol. He got in touch with Coleman again.

At about the same time, entrepreneur and pharmaceutical executive Ed Legere was looking for Woods, hoping to buy or license his ethanol patents. Legere had been running a publicly traded pharmaceutical company in California and he wanted to make a career change. "I thought bioenergy was going to be the next big growth area in biotech," he recalls, and he wanted to get in at the beginning. He had researched all kinds of feedstock. "I looked at everything, and there was a wart on every hog. Sugar and corn? Those are commodity crops. If I'm running a company whose feedstock is part of a global market, that means I'm not in control of my manufactur-

---

*Others counter that ethanol producers use only the starch from corn; the valuable vitamins, minerals, protein, and fiber are sold as livestock feed, and so still enter the food chain. Moreover, direct consumption of corn (on the cob, frozen, canned, or corn syrup) is a tiny portion of the market for corn, and foods made from corn are far more affected by processing and transportation costs, retail markups, brokerage fees, and finance costs than the price of the underlying commodity. They point to the fact that corn prices have dropped substantially since 2014, but there has been no corresponding drop in corn-based food prices.

ing process. I'd seen how that works: corn goes up to six dollars a bushel and, all of a sudden, ethanol manufacturers are hammered."

In his research he had come across Coleman's paper and Woods's patents. "Cyanobacteria made the most sense to me. The only serious input is carbon dioxide, and I knew there would always be plenty of that. If the facility was colocated with a cement or power plant, its waste carbon dioxide could be had for free. If not, at least carbon dioxide was cheap to transport." Or, certainly a lot cheaper to truck than corn, grasses, or waste lumber.

Woods and Legere, along with Craig R. Smith, a successful pharmaceutical executive and a serial biotech entrepreneur, started Algenol in 2006. Alejandro González Cimadevilla, a young member of the Corona beer family and a biotech investor, was the company's first major financial backer. John Coleman joined as a scientific advisor. "Oddly enough," Coleman said, "I still had in my lab freezer hundreds of vials of DNA from the organism we'd transformed. Typically, I save samples for five years, but for some reason I still had these after ten years. If I hadn't saved them, it would've taken a lot more time to get started."

The samples were helpful, but Coleman had only chosen that particular species because he'd hoped its particularly small chromosome would be easier to engineer. One of the first orders of business for Algenol, therefore, was to identify species that would be simple to manipulate and would make prolific amounts of sugars while tolerating the presence of ethanol. The organisms would have to withstand physical stress, as well. At the time, Algenol's PBRs were long, flat plastic tubes that lay horizontally on the ground, and water and cyanobacteria had to be pumped through the tubes at considerable pressure to propel them all the way to the end.

As luck would have it, Algenol soon discovered Cyanobiotech, a four-person startup working out of a basement lab in Berlin. The four young PhDs from Humboldt University were searching for cyanobacteria that make toxic compounds that might have anticancer or antibacterial properties. To that end, the partners had amassed a collection of species on scouting trips around Europe. Legere and Smith, speaking from hard-earned experience in the pharmaceutical industry of the challenges of turning a compound into a profitable drug, convinced the partners that they should screen cyanobacteria for Algenol instead.

Algenol soon had PBRs running in the lab in Berlin and at an outdoor site in Spain. With the help of timely grants from the Department of Energy and Florida's Lee County, the company bought a site in Fort Myers in 2010. There, the pace of discovery and innovation quickened. Cyanobiotech (which Algenol had by then purchased) whittled down a long list of contenders to a dozen and successfully inserted yeast genes in many of them. By 2010, Algenol was producing ethanol on a demonstration scale at a rate of more than 2,500 gallons per year per acre of barren scrub in Florida—six times the amount of ethanol a Kansas corn farmer could produce on an acre of farmland.

That was impressive, but Woods had calculated that it would take more than six thousand gallons per acre to reach his goal of selling ethanol, without subsidies, for under $1.30 per gallon. The problem was his scientists seemed to have hit the physiological limit of cyanobacteria's ability to produce ethanol. If the organisms diverted any more of their sugars into making ethanol, they wouldn't have enough left for basic functions. In essence, they would die trying.

Woods's last hope resided in one wild cyanobacterium, codenamed 171. It stood out above all the others as an efficient sugar producer that was also extremely tolerant of heat, high ethanol concentrations, and mechanical stress. "This bug," Woods said, "was stunning. We had been trying to transform it but had made no progress at all. Its very hardiness made it incredibly difficult to engineer. Holy crap, it was a tough bug! Really, really tough." Nonetheless, Algenol's microbiologists continued to work at it.

By the middle of 2011, Algenol was running low on cash. In the nick of time, Woods closed a $100 million licensing and investment deal with Reliance Industries, the second-largest company in India and also the owner of the world's largest petroleum refinery. Refineries produce tons of concentrated waste carbon dioxide as flue gas, and Reliance was intrigued by the prospect of turning its CO_2 into marketable ethanol. Algenol set up a pilot facility at Reliance's bayside refinery in Jamnagar, India.

Even better, not long after Reliance invested, one of Algenol's young microbiologists finally broke into 171's chromosome and successfully inserted the ethanol-producing genes. Ethanol production more than doubled. Under ideal conditions, the engineered 171 could produce as much as six thousand gallons of ethanol per acre, but not on average. It still wasn't enough.

Woods became convinced that the solution lay in the design of the PBRs. They needed a radical reengineering, he decided, and in February of 2012, he put together a new team to explore a wide range of alternatives. Algenol had been using long bags lying on the ground, assuming that maximizing cyanobacteria's exposure to sunlight would be best. But after eleven months of intense experimentation, to everyone's surprise, the data showed that the best vertical system produced eight times more ethanol than the best horizontal system. With the new PBRs, Woods said, "Suddenly, we were hitting nine thousand gallons per acre." Improvements in ethanol processing and the onsite manufacture of the PBRs further reduced costs. The company was ready to scale up to make commercial quantities.*

Many biofuel startups fail at the expansion stage. The biology, chemistry, and physics behind a venture like this change when volumes increase dramatically. For Algenol, though, the transition was simple. Algenol's approach is modular: A certain number and configuration of the plastic pouches maximizes productivity and minimizes expense. To expand production, Algenol just builds and conjoins another identical set of PBRs, much as bees add cells to

*With any live bioengineered organism there is an environmental question: Could accidental release threaten competing species and upset existing ecosystems? The EPA confirmed that Algenol's organisms are not viable outside the photobioreactors. They divert so much of their energy into making ethanol, they can no longer effectively carry out all the other tasks that wild cyanobacteria do, ensuring that they cannot survive in the wild. Now, Harvard scientists Pamela Silver and James Collins have engineered "kill switch" genes in bacteria that make them dependent on certain molecules that do not naturally occur but are added to the water in PBRs or ponds. Without those molecules, the bacteria die.

their honeycombs. Algenol's ethanol production is limited only by the availability of sunny, warm acreage with access to salty or briny water.

It takes one acre of well-fertilized, extensively watered, arable land to produce about four hundred gallons of corn ethanol per year. Algenol uses just one acre of nonarable land, no fresh water, minor amounts of fertilizers, and potentially free carbon dioxide to produce, according to Woods, about nine thousand gallons of algae ethanol. Instead of turning fresh water and corn into fuel, it turns saltwater and a greenhouse gas into fuel, with fertilizer and fresh water left over. It all sounds great. So why, I ask Woods, isn't a line forming out the door? The answer I got: Just wait!

I next saw Woods at a conference in the summer of 2015 where he was delivering the keynote. Speaking at a lectern, he highlighted Algenol's recent accomplishments and then joined other key speakers seated on the dais for a panel discussion. With a relaxed demeanor, he radiated confidence and dominated the conversation. He was clearly a man in the homestretch, ahead by lengths, sure to be the winner, generously sharing his experience and advice. When I met up with him later in the day, he told me that Reliance would be expanding the Jamnagar facility to increase its production tenfold, and an energy company in southern China was signing on to develop algae fuel projects. He hinted of potential investors in the Middle East and a new facility in central Florida that would employ a hundred people and produce eighteen million gallons of ethanol per year. In September, I read that Algenol had signed a deal with Protec Fuel Management to market and distribute the ethanol produced at the Fort Myers facility. I was buoyed by the company's impending commercial success, especially in the wake of the demise of algae oil companies. Algae ethanol was on the way to becoming a large-scale, environmentally sound transportation fuel.

Then, on the morning of October 26, 2015, I was at my computer when an alert I'd set up for Algenol popped up: "Shocker! Algenol Biotech LLC ('Algenol') announced today that its Board of Directors has accepted the resignation of CEO, Paul Woods." The company, the message continued, had fired dozens of its more than 170 employees, and would "pivot to water treatment and carbon capture now, and maybe fuels later."

I was indeed shocked. What happened? Had Woods been less than honest about Algenol's production? Were there financial problems he hadn't disclosed? Had Woods been selling green snake oil? How could a company that seemed so close to success fail so suddenly?

For a year, I could find out nothing more. Algenol, which under Woods issued regular press releases, went silent. But in early 2018, I spoke with Jacques Beaudry-Losique, the chief business officer at Algenol, and asked him those questions. "The nine thousand gallons per acre Paul talked about was accurate," he said, "but only at its best, not on average through the year. And Paul didn't include ongoing capital costs — like periodic replacement of worn-out PBRs — to our cost per gallon." Once these costs were incorporated into the company's bottom line, the cost of ethanol was too high. At this writing, the price of a gallon of corn ethanol is $1.50. That's a rock-bottom price. Algae ethanol can't match it.

Algenol's board of directors is changing the company's focus to high-value products that generate financial returns more quickly. The company now has fourteen acres of photobioreactors at its Fort Myers site and is plunging into growing spirulina for its phycocyanin blue pigment. Beaudry-Losique explained, "We have a new strain that grows like gangbusters in the photobioreactors. Its productivity is two to three times better than open pond spirulina. And its market price is one hundred seventy dollars per kilogram as opposed to a few dollars per kilogram for algae ethanol."

Algenol is betting that spirulina will be the engine of its success, now as a colorant and later as edible protein. To take that next step, engineers have had to find a way to remove the bitter taste of processed spirulina; now they want to sell a colorless, tasteless spirulina as a protein supplement and perhaps as an ingredient in synthetic meats. There is still hope for fuel, Beaudry-Losique tells me, but it all depends on market price of fossil oil. At the moment, algae ethanol, like algae oil, simply can't compete.

8

The Future of Algae Fuel

The future of algae fuel lies in advanced technology and bioengineering. Algae are engines that turn carbon dioxide and water into a huge variety of organic compounds under an enormous range of conditions. We need to optimize the right species to manufacture just one product—fuel—under just a few conditions of our choosing. Above all, it will take genetic modification, but also new laboratory techniques, perfected raceways or photobioreactors, better harvesters, and more efficient conversion technologies. The good news is that engineers have ever more sophisticated tools to make these improvements.

Until recently, testing for ideal algae strains took massive amounts of time. First, researchers grew a number of subject species in flasks on a shaker platform. Next, they grew out the most productive ones in carboys and then transferred those into bathtub-size test ponds or small photobioreactors. It took months of experiments to determine a strain's ability to accumulate oil, and even when one of them performed well in a limited outdoor test, long-run success was not assured. Questions always remained: Could changes in the pH of the water or a different regime of nutrients improve oil accumulation? What would happen as day length and temperature changed seasonally? Would a strain tolerate turbidity? Would flocculation or microwave treatment be a more efficient way to dewater? With a limited number of test ponds, discovering the optimal growing conditions for hundreds of strains would take years.

From 2010 to 2013, a government-funded consortium of thirty-nine public organizations, national labs, universities, and businesses called the National Alliance for Advanced Biofuel and Bioproducts (NAABB) undertook a wide variety of research projects. With the support of NAABB, David Kramer, Hannah Distinguished Professor in Photosynthesis and Bioenergetics at Michigan State University, invented a laboratory device he called the "environmental photobioreactor," or ePBR. The device is about the size and shape of a ten-cup coffeemaker. Researchers can mimic any climate and latitude in an ePBR to create nearly any combination of light, temperature, salinity, pH, oxygen and carbon dioxide concentration, water turbulence, nutrient levels, and other outdoor conditions. Want to know what would happen to algae when winds blow sand into ponds in West Texas or rains dilute saltwater ponds in Florida? Just tap in the changes. At only $10,000 per unit (peanuts compared to time-consuming pond trials), fast parallel testing suddenly became affordable.

But which of the tens of thousands of microalgae species — and the many millions of varieties that have arisen naturally or can be bioengineered — should go into an ePBR? Researchers at Cornell and Texas A&M invented a technology, informally known as a photobioreactor on a chip and officially termed a high-throughput droplet microfluidics screening platform, that may go a long way to speeding up the search for promising candidates. To construct a PBR on a chip, a single alga is contained in a micron-size water droplet, itself enclosed inside a nearly equally small droplet of oil. Then, millions of droplets are packed onto a chip about the size of a quarter. When exposed to light, the algae multiply in their minute containers. Researchers then analyze their growth rates and lipid production for further investigation. The device can screen millions of cells in a short time — a significant advance in the search for extraordinary algae that will revolutionize biofuel.

NAABB-funded engineers also went to work on optimizing the design of raceway ponds to solve two major problems. For one, while paddlewheels thoroughly mix the pond water and algae in their immediate vicinity, their force wanes with distance, which means a portion of the algae doesn't get maximum exposure to light. Second, algae's metabolism slows down as the water cools, so even in the warm climates of the southern United States they're less productive making sugars at night.

In 2010, engineers at the University of Arizona patented a system to address these pond design problems. Their Algae Raceway Integrated Design (ARID) system is composed of a series of shallow rectangular basins, each one sloping ever so slightly downhill to the next. During the day, water flows gradually from the highest to the lowest basin, where a pump—which uses less energy than a paddlewheel—transports the water back to the top. At night, the ARID ponds have another trick. Just below the lowest basin lies a deep and narrow trench, and after the sun sets, all the pond's water is allowed to flow into the trench. Because the ground has an insulating effect and the surface area of the trench is small, the water loses less heat overnight and warms up more quickly in the morning than it would in a standard raceway. The algae get an earlier start on full-bore photosynthesis the next morning, as well as a head start on growth in spring and an extension of their growth period in the fall. And ARID reduces operating costs by 40 percent.

These new technologies and others, including major improvements to HTL, reduce the cost of growing algae, boost their production, or both. When NAABB closed out its work, it calculated that if growers were to implement all their suggested improvements, they could price algae gasoline as low as $7.50 per gallon. While that's a massive drop from the $240 per gallon price projected by the Aquatic Species Program in 1995 and within yodeling distance of the price of fossil fuel, it's still not nearly close enough. Nonetheless, algae oil technology is still far from mature, and there are many avenues of innovation that can make it more competitive.

One way to improve algae's productivity would be to increase the amount of light they capture and therefore the amount of energy they channel into creating lipids. Photosynthetic organisms—algae and plants—have "antennas" in their chloroplasts that capture different wavelengths of light. Algae antennas use only about 25 percent of the light that falls on them, and they have to dissipate the remaining energy as heat and fluorescence, or else they suffer physical damage. Why would an alga build such dangerously large antennas? Pure defense: Any light its antennas capture will be denied to its competitors, which might otherwise grow larger and shade it. It's the same evolutionary path that led some land plants to become trees. Both spend energy on unproductive structures (large antennas in algae and non-photosynthetic tissues in trees), in order to avoid being overshadowed by their neighbors.

If you're an alga, it's all very well and good for you to maximize the size of your antennas. But if you're a human intent on producing algae oil economically, those large antennas are a problem. Overendowed algae limit the collective's ability to grow and produce oils. It's a marine version of the tragedy of the commons: what is optimal for the individual is bad for the group (at least, for our purposes). To solve this problem, Dr. Richard Sayre, senior research scientist at the Los Alamos National Laboratory, engineered strains of *Chlamydomonas* to have an optimal antenna size. A group of the optimized algae accumulate five times as much oil as their wild counterparts – a remarkable improvement.

Sayre's work is only one example of how bioengineering could change the future of algae oil. Scientists at the Danforth Plant Science Center at the US Department of Agriculture are investigating another avenue to increase algae's lipids. In 2016, Dr. James Umen and his USDA colleagues made an unexpected discovery while investigating complex protein signaling systems in algae. They created a *Chlamydomonas* variant that has a mutated *VIP* (vasoactive intestinal peptide) gene that, by interfering in protein signaling, enables the cell to make extra oil compared to unmutated cells. Moreover, despite channeling more energy into lipids, the mutated cells reproduce at nearly the usual pace. Better yet, when starved of nitrogen, the mutants go into higher gear and produce oil at nearly double the normal rate.

The *VIP* gene is far from the only one that can be altered. Scientists funded by NAABB sequenced the nuclear and chloroplast genomes of eight top algae strains and identified at least fifty genes that influence oil production. Today, with the cost of sequencing a genome down to $1,000 and new techniques (like insertional mutagenesis that generates a stream of randomly altered algae), we're more likely to discover a new algae gusher.

Genetic engineers have only just begun to explore what can be done with new gene-editing tools like CRISPR/Cas9 that they can use to purposefully and precisely alter a genome. Thanks to the technique, a joint venture between Synthetic Genomics and Exxon-Mobil recently reported a breakthrough. Their bioengineers have been focusing their efforts on *Nannochloropsis gaditana,* an organism that naturally produces high levels of oils when starved of nutrients. By sequencing *N. gaditana*'s active genes just as starvation jolted the algae into oil overdrive, they were able to isolate a gene — *ZynCys* — that fine-tunes the cell's conversion of carbon to oil. By modulating the expression of that gene, their engineered algae produced twice as much oil, yet continued to grow at normal rates.

Other researchers at the J. Craig Venter Institute in Maryland and Waseda University in Tokyo are tinkering with genes that control algae's biological clock. Algae have a natural circadian rhythm, just as we do, that puts them in sleep mode in darkness. By engineering that gene, algae can be made to think it is daylight all the time. Exposed to artificial light, they keep producing oil around the clock.

These particular genetic tweaks may or may not have a commercial future, but they are examples of the breakthroughs that are now possible. Their impact on algae oil production is in the early stages, and there is much promise.

There has been progress, too, on harvesting and transforming algae. Global Algae Innovations (GAI), a company headquartered in San Diego with raceway ponds on the island of Kauai in Hawaii, takes carbon dioxide from a neighboring power plant to grow its crop. GAI's patented Zobi Harvesters, which employ high-tech membranes, filter algae from pond water in a way that uses thirty times less energy than other processes, a major advance in reducing costs.

It's still early days for hydrothermal liquefaction (HTL). Dr. Phillip Savage, head of the department of chemical engineering at Pennsylvania State University, and his colleagues have reduced the time it takes to transform algae into oil from sixty minutes to sixty seconds by using a new process, fast hydrothermal liquefaction or FHTL, that employs fluctuating rather than constant temperatures within the chemical reactor. "Algae oil is all a matter of economics," he explains, "and FHTL allows us to reduce the size of the reactor and take less time to make the conversion, which drives both capital and operating expenses down."

Researchers at the US National Renewable Energy Laboratory in Golden, Colorado, are exploring other processing options. In a 2016 paper, Tao Dong and his colleagues demonstrated that by first treating algae biomass with a dilute acid, they were able to capture its sugars to ferment into ethanol, more easily extract the lipids, and grab the proteins for high-value products. Combined Algal Processing, they report, can reduce the cost of biofuel by nearly a dollar per gallon.

Adding carbon dioxide to ponds is essential to maximizing algae growth, but it's a significant production expense. A source of free waste carbon dioxide is ideal, but it's hard to find inexpensive land in a sunny, warm location that also happens to be right next to a CO_2-emitting cement factory or power plant. If only we could snatch the gas from ambient air and concentrate it anywhere . . .

Klaus Lackner, physicist and director of the Center for Negative Carbon Emissions at Arizona State University, and his colleagues have been working to do exactly that. Lackner has invented a passive air filter made from a commercially available white plastic resin that naturally absorbs carbon dioxide. When air passes over a dry filter, the resin absorbs the gas; when wetted, the resin releases CO_2 as a stream that is a hundred times more concentrated than it is in ambient air. The filters can be endlessly reused. Lackner would like to see the harvested carbon permanently sequestered by reacting it with certain common rock types, but it could also be fed to algae. At the moment, his technology is operating only at a demonstration scale, but it's definitely intriguing.

So, taking all these promising technologies into account, how realistic are hopes for a competitively priced algae fuel? I asked Steve

Mayfield, a professor of biology and an algae geneticist at the University of California, San Diego, and a founder of Sapphire, what it would take. His answer was prompt: ten years and ten billion dollars. My take? I've been following this sector long enough to think that it will take at least twice as long and multiples of that amount.

Discouraging? Yes and no. Companies, universities, and research organizations have been developing the technology for finding and extracting fossil petroleum and turning it into usable products for a hundred and fifty years. There are tens of thousands of petroleum engineers working today, and in 2015 more than eleven thousand students were enrolled in undergraduate petroleum engineering programs in the US alone. On the other hand, serious research into algae oil only got underway in 2008, and today just a handful of students are enrolled in algae oil engineering courses.

Worldwide, the fossil fuel industry's annual revenues are measured in the trillions of dollars. Combined profits of the six largest oil companies come in around $50 billion per year, which means a company like ExxonMobil can easily commit $1 billion a year to petroleum research and development. Fossil fuel companies have also long had government support in the form of depletion allowances, write-offs for drilling expenses, and a tax break for domestic manufacturing to the tune of nearly $5 billion per year. Moreover, we have never demanded that either fossil fuel companies or the beneficiaries of the fuels (which would be nearly everyone) cover the damages of dumping carbon dioxide into our shared atmosphere. The success of modern economies owes an incalculable debt to fossil fuels, and, until recently, no one understood how swiftly and dramatically manmade carbon dioxide emissions have been changing the climate. But we are now aware of the dire consequences of our reliance on fossil fuels.

No one knows exactly what the financial cost of our carbon dioxide dumping will be, but it is already evident and rising. NOAA estimated the total cost of the sixteen separate billion-dollar weather events of 2017 in the US at a record $306 billion. According to a 2017 article in the *Harvard Business Review,* the share of economic output at risk from climate change — hurricanes, flooding, and air pollution shutdowns; civil unrest and destabilizing migrations within and between countries caused by drought, the death of coral reefs, and the decline of wild fish populations; vector-borne diseases and pollution-caused asthma and lung cancer; firefighting, dike building, and demand for more air-conditioning, as well as other impacts — will put $2.2 trillion in US economic output at risk by 2025. This figure is an underestimate. A July 2018 report by the UN Intergovernmental Panel on Climate Change (IPCC), based on 6,000 international studies, projects that global temperatures will have risen 2.7 degrees Fahrenheit (1.5 degrees Celsius) above preindustrial levels by 2040, and that the economic cost will be $54 trillion by that date. If the temperature rises by 3.6 degrees Fahrenheit (2 degrees Celsius), as it most likely will, the cost will be $69 trillion.

So if it takes $10 billion — or $20 billion or $30 billion — to get to algae oil, not only is the price small compared to what has been spent on fossil fuel development, it is a drop in the bucket compared to the trillions of dollars we will lose to continued global warming. It took far greater subsidies for wind and solar power to become affordable ways of generating electricity. The carbon dioxide emitted by burning fossil fuel in vehicles is 14 percent of global carbon emissions. Spend $30 billion for a fuel that might eliminate even a portion of those emissions? It would be a bargain.

But maybe you're thinking: Won't we all be driving electric cars soon? Why care, then, about algae fuel?

It's true that in 2008, when entrepreneurs first invested in algae fuel, electric cars were virtually nonexistent. But many electric cars *do* create carbon dioxide emissions, just at the power plant instead of the tailpipe. Electric cars in the American South and Midwest, China, and India, for example, run on electricity generated by power plants burning, primarily, coal. And while electricity generated by natural gas is 50 percent less polluting than coal, it still produces carbon dioxide on a vast scale. Furthermore, an analysis recently published in *Science* suggests that natural gas fields and pipelines leak far more substantial amounts of methane than official estimates state. Although methane disappears from the atmosphere (through chemical reaction) more quickly than carbon dioxide does, its greenhouse effect is seventy times more powerful over the short term. That means even minor methane leaks have a major immediate climate impact: a leakage rate of greater than about 3 percent puts natural gas on par with coal. So, only if electric cars run on power generated by renewable or nuclear sources are they a substantial improvement over gasoline.

Even if we expect a huge increase in the number of electric vehicles on the roads and make the Pollyannaish assumption that they will all run on electricity generated by solar, wind, hydro, or nuclear power, consider this: According to the Air Transport Action Group, an industry organization, about 12 percent of the CO_2 emitted by global transportation comes from jets. If the global aviation sector were a country, according to *Scientific American*, it would be the seventh-highest producer of carbon dioxide emissions in the world, thanks to the 90 billion gallons of fuel that jets burn annually. Moreover, global air travel is growing, increasing by more than 5 percent each year. Then there's the shipping sector, which is responsible for another 3 to 4 percent of the world's carbon dioxide emissions. Neither large jets nor cargo ships will be moving around the globe on batteries, but algae oil makes an excellent replacement for jet fuel and bunker oil. If jets and ships alone were running on algae by mid-century, I would call it a great triumph.*

We're far from achieving even that limited goal, but I keep this in mind: The first companies to produce *any* algae oil did so only about ten years ago, just yesterday in the long process of developing a new energy source. Success will come with advances — ad-

---

*How much land would we need to replace the 17 billion gallons of fossil jet fuel the US burns each year with algae fuel? Assuming we can grow 10,000 gallons of fuel per acre, we'd need 1.7 million acres, or an area a little larger than Delaware. To replace the roughly 175 billion gallons of fossil gasoline and diesel fuel, we'd need an area the size of West Virginia. If this sounds like a lot, consider that about 20 million acres of *arable* land in the US is planted in corn for ethanol. The land required for algae production would be arid or semiarid; the American Southwest has sufficient land to accommodate production.

vances within our technical reach — in genetics, cultivation, and biomass conversion. Capturing algae's mineral components to sell or recycle will help reduce price, too.

All this takes money. If fossil oil were $100 per barrel or greater (with certainty that it would remain so), we would be on our way to making the necessary investments. But the price of oil today is considerably below that mark, and there is no company — aside from GAI and Exxon — committing funds to algae fuel research.

Government and academic support is critical; ASP and NAABB funded major scientific advances. Even when oil prices are high, entrepreneurs, driven by an urgent need to start production to generate revenue, simply can't take the time to investigate the full range of species or production options. Today, support for the sector comes through the DOE's Bioenergy Technology Office and its Advanced Research Projects Agency–Energy, which have been funding a portfolio of small projects at levels of roughly $40 million (a level that the Trump administration threatens to cut drastically). But we should do more. It's worth remembering that solar and wind power, now economical and increasingly important sources of energy, benefited from years of major government subsidies.

The economic cost of global warming, not to mention the social and political price, is rising daily. There's no question that the politics of carbon emissions regulations — whether a tax (with redistributed revenues), cap-and-trade, or other approaches — are difficult, as are the politics of government support of research. Eventually, the pain of climate change will become great enough to overcome these difficulties. But time is wasting and CO_2 levels are increasing. The fact is, we can make transportation fuel from algae. It's a question — and granted, a very difficult question — of reducing the cost of its production while gradually raising the price of fossil fuels to reflect their real cost.

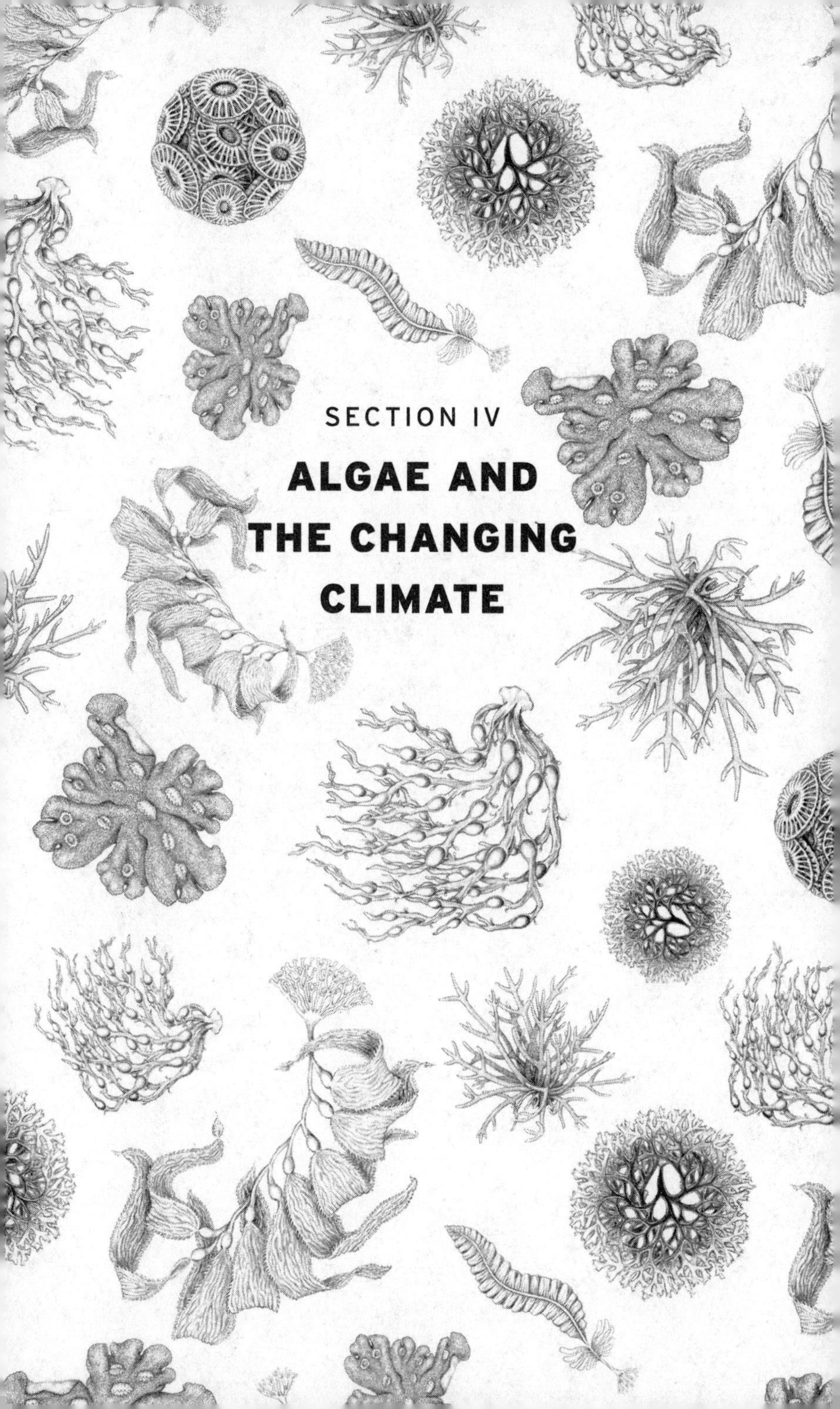

SECTION IV

ALGAE AND THE CHANGING CLIMATE

1

Gadzoox

I am standing on the very edge of the stern of the dive boat, with the blades of my scuba fins cantilevered over the dazzling aquamarine waters off the coast of Grand Bahama Island. There is a warm but brisk breeze this morning, and small whitecaps make the deck bounce and swivel, so whether it is anxiety or the wave action that is sending my stomach sluicing around my interior, I can't say for sure. Four things are certain, however. One, I must complete this, the first of four open-water dives I need to complete my scuba certification. Two, there are eight other divers in my group waiting — impatiently, I'm sure — to follow me into the ocean. Three, under the tropical sun I am cooking inside my full wetsuit, hood, buoyancy vest, booties, and fins. And last, the thirty-pound compressed-air tank on my back and the fourteen-pound weight belt around my waist make standing upright a serious challenge.

Nonetheless, I linger. I'm scared.

All the other divers are young couples getting their certification here before continuing to dive on their vacations or honeymoons. I have come to the reef not for pleasure, but for research. Although coral reefs cover only about 1 percent of the ocean floor, a third of all marine fish species spend at least part of their lives here. Juvenile sea animals grow up among the corals' nooks and crannies safe — well, safer — from predators. The corals also function as submerged breakwaters, dissipating the energy of waves by as much as 97 percent and defending the shoreline from erosion. As coral reefs disappear, islands gradually follow.

The economic import of reefs is enormous. According to the World Wildlife Fund, the planet's coral reefs generate about $30 billion of goods and services per year. The UN Food and Agriculture Organization reports that 17 percent of the world's protein comes from reefs. One billion people depend to some degree on them for food, protection, or employment. In Indonesia, the country's 250 million inhabitants depend on reef fish for the majority of their protein. And then there's the immeasurable psychic value of their glorious Technicolor beauty. And none of it is possible without algae.

Which is why I'm here, scaring myself: All the beauty, the teeming marine life, the revenues and employment, all of it depends on the algae that live inside corals. All the fish, mollusks, anemones, lobsters, and other reef animals sustain themselves by eating algae, directly or indirectly. Algae also literally cement the reef together. Remove the algae, and the reef disintegrates and all of its inhabitants vanish. Even the soft white sand vanishes. Almost all of it is the calcium carbonate remains of a lovely green seaweed, *Halimeda*. Without algae, tropical oceans would be watery deserts.

"Don't look down," our dive instructor, cupid-cute, curly-haired Aaron from South Africa, urges me. "Keep your eyes on the horizon and just take a really big step out."

There is nothing for it but to go. I hold my mask and regulator to my face with my right hand, cover the buckle of my weight belt with my left, and stride into the void. I hit the water, plunge beneath the surface, and, in a crescendo of bubbles, immediately pop to the surface. My inflatable vest keeps my head well above the chop, but I continue to breathe through the regulator. My heart is pounding, but the cool water sliding into my wetsuit is a relief. I make an OK sign with thumb and forefinger and kick away from the boat.

One by one, the other divers jump into the water, and Aaron signals that it's time to descend. One after another, my dive mates raise their inflator hoses above their heads, let the air out of their vests, and sink feetfirst beneath the waves. I do the same and gradually submerge, as I watch a stream of bubbles rise above me.

But not for long. I am perhaps six feet below the surface when I get stuck. Not only is my body not sinking, my feet are now slowly rising. I try to press them down, but it's impossible; they have a mind of their own, and they want to go up. As they rise, my torso tilts back, so I feel I'm in the first stage of a backwards somersault. My legs angle toward the surface as if drawn by a celestial magnet. I flap my arms — a bit frantically and completely uselessly — trying to remedy my predicament.

Aaron swims to my rescue. He helps me turn upright, moves some of the weights in my dive vest to my belt, and transfers an additional weight from his belt to mine. He pantomimes that I should breathe more slowly, using longer exhalations. With his hand grasping my ankle, we head downward. Halfway down, he releases me, and I continue on my own. I fall in slow motion toward the white, sandy bottom — an underwater beach — and take my place on my knees beside the others. Despite the turbulence on the surface, the water around us is motionless. The divers around me look as if they are encased in glass.

The first order of business is demonstrating, one by one, a number of scuba skills, like doffing and donning our face masks and clearing them of water, and switching to our backup respirators. Work done, Aaron points ahead and makes a dolphin motion with his hand across his chest, and we set off in a haphazard line, moving slowly. *Languid* is the word of the day. We are tourists, getting the lay of the submarine land, sliding through the reef for an hour's outing. A school of deep-blue fish nearly as round and thin as salad plates immediately joins us, gliding just out of reach, a crowd of curious fellow travelers. A saucer-size angelfish, jet black with a brilliant yellow head and tail and astonishing blue lips, crosses our path. Other fish in improbable colors and patterns — think zebra-striped — pass alongside or under us. Ahead, in deeper water, a school of hundreds — no, certainly thousands — of small, knifelike, silver swimmers flash intermittently as they turn and turn, en masse.

Leaving the white sandy bottom, we now fly over a new landscape: the coral reef itself. Corals are everywhere. An orange one that looks like a small sapling lives next to a coral the shape of a brain and the size of my kitchen table. Within a few dozen yards I see corals that look like a black fan, a green fern, a rusty iron wire that spirals six feet out of the seafloor, a candelabrum, and a group of thick orange fingers. There's a tangle of branched twigs resembling a sunburned tumbleweed, and tall corals like the underwater cousins of Texas cacti, albeit in odd colors — one red, some greenish gray, another a lovely pale yellow.

There's plenty of algae to see, too. In fact, it's growing on nearly every rock and many corals, dead and alive. I see what look like boulders below me, and I breathe out slowly, emptying my lungs in order to sink and observe them more closely. The boulders turn out to be corals covered in rough patches of pink and orange and red. It's as if a gang of six-year-olds went wild with buckets of colored stucco.

The stucco patches are crustose coralline algae, members of the red algae group, which often appear in rosy hues, but also in purple, blue, and gray, and inhabit every ocean. They grow slowly, advancing across and encrusting the surfaces of rock, shells, and other corals. With tough deposits of calcium carbonate in their cell walls, they are a kind of living limestone. Over time, layers extend over layers, growing ever thicker, transforming heterogeneous bits and pieces into a rock-solid reef.

Crustose coralline algae not only create the reef, they help populate it. Drifting larvae of bottom-dwelling creatures like oysters, snails, and mussels must find a solid surface to attach to in order to metamorphose into their adult forms. These algae produce chemical compounds that cue the larvae to attach, like real estate agents baking cookies just before an open house. After the larvae settle in, they are sheltered from predators amid the corallines' nubby, knobby, and sometimes deeply crenelated surfaces. Their home is indeed a castle.

Turf algae (a general term for all kinds of low-growing algae), like the kind that covered the bottom of my pond, would love to anchor themselves on top of crustose corallines. But the turf algae would deprive the corallines of sunlight, so they periodically slough off cells from their surface, ridding themselves of these unwanted relatives. Fish unintentionally help out, too. As I watch, little swimmers — some black and blue, others as bright yellow as goldfinches — dart at the coral boulders, picking away at the fuzz of turf algae like hungry birds pecking for crumbs on a sidewalk. I hover, entranced, and lose track of time.

I am startled by a tug on the end of one of my flippers, and I twist to see who — or what — is there. It's Aaron, holding the air pressure gauge attached to his vest by a cord, and pointing to it. He is asking me how much air remains in my tank. I was supposed to be checking my own gauge every ten minutes, but I've forgotten. I fumble for the gauge at my waist and squint. It reads 1200 psi, which means I've used up more than half my air, but I'm in no danger. I put one finger on my forearm and then hold up two fingers. He signs "OK" and "wait," and then swims off to query the rest of the group. Five minutes later we regroup, reverse course, and proceed more purposefully back over the coral toward the boat. At the ladder, I slip off my fins, pass them aboard, and clamber up.

I'm ready to dive again.

Two days later, we meet for our final dive, this one just for fun and at night. I stride fearlessly off the resort's dive dock into the dark air, and surface from the pitch-black water to wait for the rest of our group to assemble. Wearing wrist-mounted dive lights, we head for a vertical reef only a ten-minute swim from the dock. Our destination is at no great depth, and a slight current means I barely have to flutter my fins. As I drift in the eerie blackness, the beam of my light reveals a different ocean. Water that seemed crystal clear during the day I now see is actually dense with tiny particles, all glowing white. It's as if I've abruptly opened a door to a shuttered and disused room, simultaneously stirring and illuminating a cloud of dust motes — but here, most of the motes are alive. They are zooplankton and the larvae of squid, eel, snails, barnacles, and mussels. They live in uncountable numbers, all swimming along hoping to bump into cyanobacteria and microalgae they can snag and eat.

We approach the reef, and at first I am transfixed by the way my light picks brilliant fish out of the blackness. Aim almost anywhere and a flash of blue, yellow, or silver pops into view and then quickly disappears. But it is the corals that shock me. How strange: They are blooming. The stems of shrublike corals wear the daintiest of daisylike flowers. Flat, nickel-size yellow blossoms blanket a rock. Salmon-colored petals decorate a twiggy coral. The more closely I look, the more flowers I see. Many are the size of a fingernail, but some are no bigger than BBs and a few are as big as Ping-Pong balls. Under my spotlight, the reef is a flower garden, blooming exuberantly, quivering, brilliant against a deep black background.

These flowers, though, are not lures for pollinators, they are predatory animals. Soft, transparent, and stationary, the animals are polyps — the living component of corals. What look like petals are the polyps' tubelike tentacles, which they thrust out to sting prey, either stunning it or killing it outright. After the attack, polyps retract their tentacles, pulling dinner into their central digestive cavity. Dinner is usually zooplankton, but larger polyps can snag a juvenile fish. Come dawn, the polyps retreat into their stony exoskeletons, protective structures they are constantly adding to with secretions of calcium carbonate.

Coral polyps

For all their effort, polyps manage to collect only about 10 percent of their nutrients from their spectacular nightly hunts. It is microalgae, living unseen inside the tissues of the polyps' tentacles and digestive cavity, that are the real workhorses and provide the other 90 percent. Millions of microalgae, which are known as zooxanthellae (zoox, for short), typically live in a single polyp. The zoox capture sunlight, pull carbon dioxide from the water, create sugars, and then leak most of their treasure to their host. Recently, biologists discovered that zoox are essential even to corals that live deep in the ocean where little sunlight penetrates. The deep-water corals have proteins that create a phosphorescent glow that their zoox use to create sugars.

Zoox can live independently, if more precariously, outside a host, but coral that lose their resident algae die within a couple of months, at most. Nonetheless, the relationship between zoox and polyps is not one-sided. By respiring carbon dioxide, polyps provide some of the raw material that the zoox need to create sugars, and they pass along nitrogen gleaned from their diet of zooplankton.

The zoox-coral relationship is one of the coziest and most fruitful symbioses on Earth, a model of beneficial cooperation, but zoox happily team up with other animals, too. Many sponges and anemones and some mollusks, marine worms, and jellyfish also depend on their algal residents. The bright colors of corals and these other animals come from the combination of pigments of the zoox and their hosts.

A healthy coral reef ripples with life. Crustaceans slip in and out of crevices; starfish clamber over its surface; jellyfish scoot about; schools of fish dash, flash, and pivot in perfect synchrony. Every inch of a reef's rugged surface is covered with animals and algae, all competing for territory. But the abundance of a coral reef is a paradox: The shallow, warm waters are short on nutrients, especially nitrogen. Why are there so few? The real question is why are there so many more nutrients in colder waters. The answer is location, location, location.

In temperate climates the surface waters cool in autumn, become denser, and sink. As the top layer of water descends, it forces deeper water toward the surface in a process called upwelling. The deeper waters are rich in the remains of dead plankton that have drifted down past the zone where most living creatures reside. When these nutrients rise in autumn, they refresh the marine buffet, supplying dinner to hungry microscopic creatures who then flourish — and cloud the water. But in shallow tropical zones, where water temperatures hardly vary, there's little or no seasonal upwelling, and there's less food for creatures at the bottom of the food chain. The water's astonishing clarity reflects its remarkable dearth of microscopic life.

So, what explains the abundance of animals in a tropical reef? Efficiency. Reef inhabitants capture and reuse every bit of available nutrition. Algae are eaten by zooplankton, which are eaten by fish and coral polyps, which die, decompose, and become nutrients for algae. Although some juvenile fish will grow up and swim away, and occasional violent storms, bird droppings, and runoff will bring in some new nutrients, overall, reefs are closed systems akin to giant aquariums. They rely heavily on recycling.

Reef ecosystems are fragile; the collateral effects of small changes can be dramatic. Take water temperature: In the last forty years, rising levels of greenhouse gases have increased the sea surface temperature by about 1 degree Fahrenheit (0.5 degree Celsius). That doesn't sound like much, but it's enough to cause the zoox's rate of photosynthesis to increase. More photosynthesis might sound like a good thing — more sugar and more oxygen production for the polyps, right? — but it's not. Not all oxygen molecules are created equal, and under the stress of high UV light and heat, zoox produce more superoxides, that is, oxygen molecules with a negative charge. Superoxides are highly reactive; they're molecular firecrackers. They damage chloroplasts and mitochondria, destroy lipids and ATP, and break down cell membranes. (In fact, animal immune systems use them — sparingly — to destroy bacterial and viral invaders.) When polyps sense the danger from their overactive zoox, they expel them. Because polyps are transparent and their exterior skeletons are made of white calcium carbonate, when the green zoox depart, corals turn white—hence the term *coral bleaching*.

If the warming is temporary and the water returns to its normal temperature quickly, the corals welcome back their zoox and recover. But if conditions don't change, the polyps starve to death. Episodes of coral bleaching have occurred in the past when the El Niño weather pattern warmed ocean waters, but now these periodic temperature spikes are superimposed on long-term climate warming. The corals have no time to recover, and the damage is permanent and widespread.

To add to the corals' misery, zoox are also endangered by the nitrogen from fertilizers and sewage that find their way into offshore waters. These algae have evolved over millions of years to live with a certain mixture of nitrogen, phosphorus, and other minerals. Runoff changes the blend, upsetting the accustomed balance, often with adverse effects. Researchers at the University of Southampton in England have found that too much additional nitrogen, for example, starves zoox by preventing them from taking up phosphates they need for cell function.

The nitrogen also feeds disease-causing bacteria and turf algae all too willing to blanket the surface of corals in a fuzzy green carpet. In the past, parrotfish, groupers, surgeonfish, snappers, and long-spine sea urchins efficiently mowed the turf algae from the corals as fast as they grew. But overfishing in the Caribbean has severely depleted these salutary and beautiful animals, and they no longer can keep up their coral maintenance chores. To make matters worse, thirty years ago a mysterious disease killed 90 percent of the urchins, and these slow-moving creatures, which look like pincushions bristling with long black needles, have yet to recover. As a result, turf algae are overrunning corals and shading out their zoox, which then can't generate enough sugars for themselves and their partners. A final blow: coral larvae cannot settle down on corals covered in turf algae.

On one of our dives in the Bahamas, we swam for hundreds of yards above the bleached remains of dead coral thickets. The sight was shocking: a white sand desert littered with the white and broken bones of corals. The disappearance of zoox—forced out of corals by heat, damaged by runoff, and smothered by turf algae—is creating a barren landscape. Almost all of Florida's coral reefs are gone. Sixty percent of the Caribbean's living corals have disappeared in the last thirty years, and it is likely that the rest will vanish in another twenty. Thirty-five percent of the coral on the northern and central portions of Australia's Great Barrier Reef are dead, and 93 percent of the corals left are affected to some degree. The prognosis is grim; many experts expect that coral reef ecosystems will be largely extinct by mid-century, at the latest.

2

Saving the Reefs?

Is there any way to save coral reefs? Duly certified to dive, I fly from the Bahamas to Bonaire, a Dutch island in the south Caribbean Sea, to find out. Within hours of landing, I am gliding twelve feet below the ocean's surface, just above the sandy bottom, and a few miles from the Buddy Dive Resort. I'm accompanied by Nathalia Castro, a Brazilian-born scuba instructor and dive master at the resort, and we're approaching one of the strangest underwater landscapes in the Caribbean. Ahead of us is an orchard of "trees" made out of white PVC piping. The trunk of each tree is about eight feet tall and three inches in diameter. Running through the trunk, at regular intervals, are eight branches made of narrower plastic pipes. Each tree is suspended in mid-water, tethered to the seafloor by a short chain and held upright by a line running from the tree's apex to a subsurface buoy. The trees are free to sway and turn in the current, and they do.

As if that weren't strange enough, from each branch dangles what look like a dozen funky, salmon-colored Christmas ornaments. The ornaments are pieces of branched coral, but this is not some sort of tropical underwater seasonal display. It is a coral nursery, part of the effort of the nonprofit, Florida-based Coral Restoration Foundation to rescue dying coral reefs. All the corals here are fragments of two critically endangered species: staghorn (*Acropora cervicornis*) and elkhorn (*Acropora palmata*).

Staghorn and elkhorn corals look remarkably similar to the antlers of their namesakes. Staghorns have cylindrical branches that furcate extensively and can grow larger than six feet in height and width. Elkhorns have flatter, more massive branches and grow up to twelve feet in diameter. Until the 1980s, these two species dom-

Coral tree

inated the Caribbean reef and created miles of dense underwater thickets. The thickets were the physical and ecological foundation of the reef, breaking the force of waves and creating calm zones behind them where fish, soft corals, crustaceans, anemones, mollusks, starfish, and other sea animals found food and refuge.

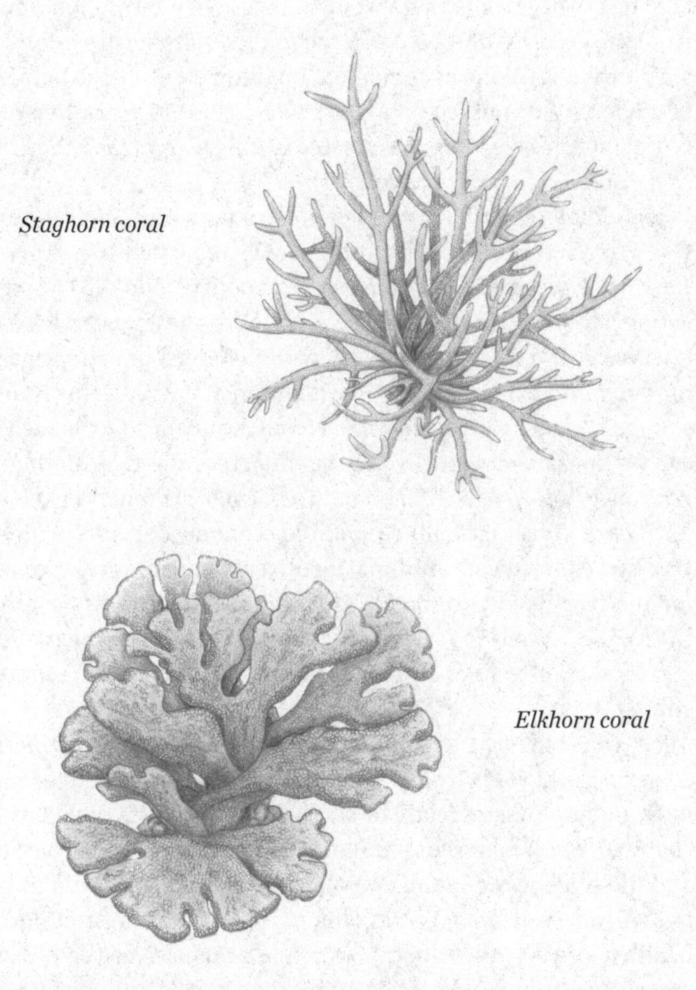

Staghorn coral

Elkhorn coral

Staghorns and elkhorns have underpinned reefs and tropical islands for many thousands of years, but the Caribbean has lost a catastrophic 95 percent of these two species. The structure and life of the reef has changed dramatically as a result, and all for the worse. In the 1970s, these corals formed a continuous and impenetrable rampart around Bonaire. The first dive shop owner on the island had to hack a channel through them so that divers could swim from the beach to deep water. Even in the 1990s, healthy stands of staghorns and elkhorns remained. But after creating the bulwarks of modern reefs in the region for at least 220,000 years, these two critical species have been decimated in only two decades. Now, just a few isolated groups survive.

Nonetheless, Nathalia and her colleagues at the Coral Restoration Foundation do not feel hopeless. It helps that they work on Bonaire: A combination of government policies and the luck of location have made the island's reefs among the best preserved in the Caribbean. For many years, the government has banned spearfishing and limited other means of fishing, and it upgraded the island's sewage system in 2012. The island itself is built of coral, made long ago by the same creatures that manufacture the reef, so there is not much soil to slip off the land to become sediment in the water. There's little agriculture, and the economy depends on diving tourism, which means minimal fertilizer runoff and great concern for the marine environment. And because Bonaire lies south of the hurricane belt, its reefs have largely been spared the monster storms that elsewhere dislodge corals and cover reefs in sand and silt.

Yet the staghorns and elkhorns are on the verge of extinction even here, which is why Nathalia and I are floating in front of a tree in the coral nursery, ready to add baby corals to its branches. By "baby corals," I mean coral fragments, each about five inches long and the thickness of a finger. Some have forked ends or a nub or two of incipient branch. Nathalia has clipped them from larger, healthy corals in the nursery, which are themselves the progeny of corals collected from different places around the Bonaire reefs. Suspended on these trees, the small corals will be protected from drifting sand and colonizing creatures, and they will face less competition for the zooplankton swimming by. They also have plenty of growing room. All these advantages mean nursery corals grow significantly faster than wild ones.

My first task today is to take up one of these amber-colored coral fragments and, using an eight-inch piece of fishing line and a little metal sleeve, put a noose around its midsection. Next, while I hold the noose in place, Nathalia uses pliers to crimp the sleeve closed, securing it. Then, I thread the long end of the noose through a set of two small holes in a plastic branch, one on the bottom and one on the top, and Nathalia crimps a sleeve onto the free end of the monofilament so it can't slip out of the holes. We'll repeat this process until we run out of coral fragments.

It sounds simple, and it would be on land. But for a novice diver, it's damnably difficult. Every breath of compressed air I take makes me rise just slightly, while every exhalation makes me sink a little. Also, the slight current this morning wants to carry me away from the tree. Another problem: I'm wearing twenty-four-inch fins and working near the seafloor. Whenever one of my fins accidentally touches the bottom, I raise a cloud of sand that obscures what we're doing, and we have to wait until the current clears the water. Nonetheless, I manage to hang all the fragments on the tree, and then help Nathalia with the weekly nursery maintenance. Using a scrap of a blue scrub pad, I scour turf algae from the tree and monofilaments and pick off parasitic snails. The turf algae are persistent and aggressive, and they have thoroughly furred the branches in the week since it was last cleaned. It's a vivid demonstration of how, with the decline of the parrotfish and other grazing animals, corals can become quickly overgrown.

Nathalia points to what looks like a fuzzy, bright-orange caterpillar inching down a filament. This is a bearded fireworm, a three-inch creature that is especially fond of dining on coral polyps. She had warned me about these: They look harmless, but if I touch their bristles, I'll get an injection of a neurotoxin that causes pain for hours and may mean a trip to the hospital. Fireworms have their place in the ecosystem, but not here in the nursery. Early on, Nathalia told me, she would cut the fireworms in half, but then she discovered that, hydralike, they turned into two individuals. She now picks them off with the needle-nose pliers she uses for cutting and crimping and puts them in a closed waste bucket. Once they're on land and exposed to air, they die.

Our next task is to gather any corals ready to be transplanted. I swim behind Nathalia to a group of trees with dangling corals that, over the course of a year, have grown to the length of my forearm and branched into spiky, complex, three-dimensional shapes. They've grown so dense they now obscure the branches they're hanging from, so the trees resemble a very peculiar species of orange conifer. Reaching into her kit of implements, bags, and bottles, Nathalia takes out a scissors and snips free a dozen large corals, keeping the filament ends for the trash. With our hands full of corals we flutter kick our way to their new, permanent home.

There are two ways to transplant corals. One is to glue them with marine epoxy to rock. This works well, but finding appropriate sites and chiseling flat places on which to set them takes time. Today, Nathalia and I use another method, tying them to rebar frames that look like ten-foot long, one-foot-high rusty staples whose ends are sunk deep into the ocean floor. My job this morning is to position the corals on the rebar in a way that maximizes the contact between the coral and the frame. This is easier work for a novice: Holding a coral to the metal, I anchor myself while Nathalia floats next to me and secures the coral with heavy-duty plastic ties. If all goes well, this new rebar frame will be obscured by corals and will resemble a thicket much like those that have disappeared.

Before we return to the surface, we tour the dozens of coral hedges at this site, one of seven sites in progress on Bonaire. Some of the hedges are covered in interlaced, spiky orange staghorns, and the rebar is completely camouflaged. I feel an outsize pride in my small contribution to restoring this beautiful world.

But will this actually work long-term?

Why should it? After all, we're still burning fossil fuels and sending greenhouse gases into the atmosphere. As much as 90 percent of the additional heat created and trapped by those added gases has been absorbed by the oceans. El Niños will continue to develop every two to seven years. The 2015 El Niño season created the greatest surface temperature increase on record—as much as 5 degrees Fahrenheit (2.8 degrees Celsius)—and lasted longer than ever before. (According to a 2014 study published in *Nature Climate Change,* the frequency of El Niños will double in the future due to rising ocean temperatures.) In warming waters, bacterial diseases will only become more virulent. Rain will continue to fall on land, washing sediment and fertilizers into the sea. Although island governments are making progress on stemming sewage discharge, the problem won't be solved for years. So how could planting corals under the same and worsening conditions succeed?

The next day I sit down with Francesca Virdis, the coordinator in Bonaire for the Coral Restoration Foundation (CRF), to talk about the future. CRF is dedicated to developing simple, inexpensive, and easily duplicated coral restoration techniques. It now has five large restoration sites in the Florida Keys, as well as the seven in Bonaire.

Virdis is thirty-nine years old, slender and dark-haired, with the refined features of a Renaissance Madonna and a lilting Italian accent. She grew up on Sardinia in the Mediterranean Sea in a family passionate about every aspect of the ocean, but particularly scuba diving. When she was ten, she started diving, too. Sardinia has no reef, she explains, but she spent her childhood exploring tunnels, caves, and shipwrecks, and watching schools of silvery barracudas and other large fish. After getting her master's degree in marine science, she went to work in Italy's oil sector, focusing on environmental safety in offshore drilling.

She was on an excellent career path, she tells me, stable and well paid, especially compared to her fellow graduates doing research. "Basically, in Italy, the career for a marine scientist is like working for free for the first ten years." Working in the oil industry, however, had its downsides, including having to deal with "a very macho environment." She missed research and had always enjoyed teaching. To satisfy the latter interest, she qualified as a dive instructor and taught scuba classes on the weekends. But after five years, she'd come to hate her job, so, she decided to take a break, become an instructor at Buddy Dive for a year, and improve her

English. She hoped that with better English she would have new career opportunities.

One year in Bonaire became two, and then three. In early 2010, the resort decided to sponsor CRF and applied for government approval to establish coral restoration sites. Virdis was involved in the project from the start, and eight years later, she is still joyfully, if not especially prosperously, involved. She now works half-time as a dive instructor at the resort and half-time as the coordinator for CRF projects in Bonaire.

The work of the foundation is predicated on several unproven but persuasive propositions. One is that any elkhorn and staghorn corals still alive after the massive bleaching and disease episodes of the last two decades must have a suite of genes that gives them a survival advantage. By taking fragments of these tough survivors, nurturing and outplanting them, Virdis and her colleagues believe they can support the reef's own regenerative power.

CRF is also betting that restoration will work by providing, in essence, a dating service. For the most part, staghorns and elkhorns reproduce asexually, by fragmentation. A branch breaks off, takes hold, and grows as a clone, allowing a coral to colonize its immediate surroundings efficiently and continuously. But there are risks to asexual reproduction, especially in a changing environment. All the cloned corals have identical genomes, and if their genes do not enable them to meet the challenges of new conditions — like polluted, warmer water — then they all die.

Fortunately, cloning is not the only way these corals reproduce. There's also the annual orgy. Once a year, and only for twenty-four to forty-eight hours, corals engage in a synchronous mass spawning. In the Caribbean, the spawning always occurs at night in August or September (it depends on the water temperature), four to six days after a full moon. When the time is right, individual staghorns, elkhorns, and dozens of other acroporid corals, all of which produce both eggs and sperm, release them simultaneously. The gametes look like tiny translucent pink balloons, and millions upon millions rise slowly, diagonally with the pull of tide or current, to the surface. This coral spawning is one of the wonders of the marine world: shine a light through the water, and you see an upside-down pink meteor shower. At the surface, wind and waves break apart the balloons, and eggs and sperm from different corals mix.

A few days later, the resulting embryos develop into swimming larvae, called planulae. The planulae then head for the bottom to attach to the substrate. The rate of implantation is low: Most of the eggs and sperm never meet, and most planulae fail to touch down in a propitious spot. Even those that successfully drop anchor are highly vulnerable to predators until they can secrete enough calcium carbonate to create a protective home. Still, this beautiful but fraught sexual process has been generating polyps with slightly different genotypes for millions of years.

Before the mid-1980s, Bonaire's reef was chockablock with staghorns and elkhorns. If you observed them closely, you could spot slight differences – say, a more vivid color or thicker branches – among clusters of corals. Their varied appearance (or phenotype) reflected slight differences in their genetic makeup (or genotype). Sometimes the dissimilarities were purely serendipitous, but in some cases, they reflected the fact that conditions – say, the clarity of the water – in one marine neighborhood were just a little different than those down the watery lane, and that the locals had evolved to prosper in their particular corner of the world. Because staghorns and elkhorns were so plentiful and lived within spawning distance of many others, their gametes would constantly mingle, generating diversity that helped them survive.

Today, however, around Bonaire and throughout the world, only small outposts of these corals remain, and the distances between them have grown so large that there is far less sexual mixing. Without human help, staghorns and elkhorns are now unable to generate the genetic diversity they desperately need to adapt to a rapidly changing environment.

This is where the Coral Reef Foundation staff and volunteers come in. They collect and tag coral fragments whose slightly differing phenotypes (or simply their remote locations) indicate they have different genotypes. After fostering the diverse fragments in the nursery, they plant them on rocks or rebar frames where their gametes are likely to cross-fertilize. The hope is that new genotypes will emerge that are better adapted to Bonaire's current conditions. Maybe, the thinking goes, the new planulae will survive because they have a mix of genes from the island's most tenacious survivors.

By keeping track of the different genotypes in their artificial thickets, Virdis and her colleagues can determine which have the greatest rate of survival at the five locations. Unfortunately, there is no correlation between genotypes that prosper in labs and those that will thrive in a particular spot in the ocean. There are too many variables in the wild — temperature, current, predator pressure, water turbidity, and more — that influence survival. Only by observing corals in their habitats are they able to see trends that will help them further select and transplant the evolutionary winners.

So, how is it going? It's too early to say for certain. Virdis says that survival rates range from 50 to 80 percent, but the degree of success depends on location.

"When a reef is already damaged," she explains, "it is harder for transplanted corals to survive. The ecosystem is out of balance, and there are too many fireworms and snails and too much algae. I've seen damaged reefs where there are thousands of fireworms because there are no crabs to eat them, and there are no crabs because there are no corals for the crabs to hide in." The first transplanted corals often suffer heavy losses and grow slowly, but these pioneers create a better ecosystem for the next generation of transplants.

For several years, Buddy Dive was the only CRF sponsor on Bonaire, but three other dive resorts have recently joined the effort. So far, the foundation has cultivated and transplanted eleven thousand corals and has ten thousand more growing in the nursery. All four resorts run coral restoration projects and offer training, and more than five hundred divers — tourists and residents — have participated. Two new sites are accessible to snorkelers.

The goal, however, is not to replant the reefs off Florida or Bonaire by hand. Thousands of acres of acroporids have been lost, and no number of volunteer divers can restore what is gone. Conservationists hope instead to boost a natural recovery and to preserve as many genetic varieties as they can before they vanish. The CRF Florida nurseries have accumulated 140 different genetic strains of staghorns, including dozens that are now extinct in the wild.

Compiling a diverse library of genotypes is critical to scientists exploring the possibilities of "assisted evolution." Marine biologists around the globe have been urgently investigating exactly what makes some corals better able to survive warming waters than others. What they've discovered is that the genetic endowment of the polyps is only part of the story. As important, it turns out, is the genetic makeup of the zoox inside them.

When we talk about zoox, we're talking primarily about members of the genus *Symbiodinium*. Until 1990, scientists believed that there was only one *Symbiodinium* species that resided in all corals. But in the 1990s, genome sequencing revealed hundreds of previously unknown species. The genus is now organized into nine clades (that is, groups of genetically similar species) lettered A through I. Each clade is further subdivided into numbered types. Until recent research on acroporids, though, no one knew if the distinctions among clades held any significance.

Acroporids generally do not tolerate water temperatures outside a narrow range, and when the temperature rises above their upper limit, they eject their zoox. In 2006, Ray Berkelmans and Madeleine J. H. van Oppen at the Australian Institute of Marine Science published a study of *Acropora millepora*, a common coral species that looks like a dense cluster of dozens of green, pink, orange, or purple fingers thoroughly pocked with holes. *Millepora* (which means "thousands of holes") are especially temperature sensitive, and many bleached and died during the 2005 El Niño. Some did survive, though, and when the scientists analyzed these unusually hardy individuals, they discovered that they no longer harbored the usual clade C *Symbiodinium*, but now had clade D symbionts.

Berkelmans and van Oppen were the first to show that corals' survival depends on the particular strain of zoox that inhabits them. The discovery led scientists to wonder: If all corals had access to clade D zoox, would they better survive in warmer water? Maybe introducing clade D zoox to warming waters could save the reefs, or at least buy time to figure out a more robust solution.

I call Dr. Ruth Gates, who heads a lab at the Hawaii Institute of Marine Biology at the University of Hawaii and is also interested in this question. A biologist who focuses on how corals cope with environmental change, she collaborates frequently with Dr. van Oppen. "I'm interested not just in investigating who is paired with whom," she tells me, "but what is happening when the symbionts operate together. Are all interactions and partnerships the same, or are some higher quality than others? A lot of people have been saying, let's get on with distributing clade D. I say, let's pause for a moment."

Research published in 2008 and 2009 concluded that although clade D *Symbiodinium* are heat tolerant, they produce fewer sugars, and so the corals harboring them don't grow as well. It's no surprise, then, that researchers also discovered that when waters cool to their normal temperature, corals eject their clade D inhabitants and take up again with clade C partners, who they can count on to produce more sugars for themselves, and thus more energy for their partners. Clade D symbionts, as it turns out, are foul-weather friends, acceptable only in an emergency.

The relationship between polyps and zoox is complex and deeply intimate. The entire biology of both organisms is tuned to the provision of materials flowing from one to the other. "I think a lot about how you evolve symbiosis," Gates says. "Almost all successful symbionts have evolved from a parasitic relationship. Over time, that relationship becomes more and more interactive and can evolve over millions of years, first to a sort of peaceful coexistence and ultimately to a mutualism where both parties benefit from the association." The genetic differences between the clades, although they may seem slight, turn out to be significant. "My position is we may be able to adjust symbiosis in threatened corals, but in very, very subtle ways and using symbionts very, very closely related to the ones intrinsically found. Swapping D for C is going too far."

Gates and her colleagues are studying a type of coral called

Porites, which is among the hardiest corals in the world's oceans, and the most resilient in the face of warming waters. *Porites* species are sometimes called Jeweled Finger corals for their shape and range of hues. They grow more slowly than acroporids but can become the size of boulders and live for many hundreds of years. Gates is currently trying to populate other corals with the *Symbiodinium* (clade C, variety 15) that inhabit *Porites*. She's experimenting with juvenile corals because, she notes, "babies are more entrepreneurial about whom they'll associate with." Should other corals take up C15 and prove themselves better able to handle higher temperatures, it may be possible to seed a reef with lab-bred juveniles or grow them up and outplant them, creating a new reef better suited for a changing climate.

But perhaps there's still another way to save the reefs. Gates and others find that wild corals that manage to survive a bout of excessively warm waters are more resilient during the next warming episode. They posit that the expression of certain genes in these corals has changed. Gene expression – that is, when and for how long a gene is activated – plays a role in how any individual looks or interacts with its environment and how it can be influenced by external circumstances. Scientists call these epigenetic factors. Sometimes, the traits an individual develops through epigenetics are passed on to the next generation.

Gates and van Oppen have been running experiments in their respective labs, gradually subjecting corals to ever higher water temperatures and pH levels, two stressors that corals face in our era of climate change, to try to build up their tolerance of these conditions. Not all the corals survive the treatment, but some do, and the question is how they do it. One possibility is that genes related to heat tolerance have been unusually expressed, and so the two scientists are now exploring whether they can pass this fortitude to their offspring. If so, we may be able to seed reefs with their gametes or larvae, which, in theory, would be more fit for the future.

The scientists in Gates's and van Oppen's labs are providing more precise scientific information to organizations like CRF. They believe that restoration professionals, armed with knowledge of how to nurture young or small corals and how to outplant them, are essential. Saving genotypes makes sense, too. But a better understanding of coral biology is also critically important. "Our goal," Gates tells me, "is to close the gap between the very rapid rate of change in temperature and acidification of the oceans and the natural rate of coral evolution. The complexity of the relationship between symbionts is enormous. It's not only the DNA of the polyps, the clade and type of *Symbiodinium,* and epigenetic factors, but coral microbiomes are involved, too. Our own health is intimately linked with microbes that inhabit our gut, skin, and other organs. No doubt, coral's microbiomes are important to their health, too. We're trying to understand and possibly manipulate all these variables as quickly as possible because we don't have a lot of time."

Gates recognizes that her work is a stopgap measure. "If we don't successfully address carbon dioxide emissions," she adds, "nothing we do for the corals will have an impact. There are only so many degrees of temperature rise before the ocean becomes unequivocally lethal."

After my conversation with Gates, I call Virdis to get her take on how coral biologists' work may impact reef restoration. She is hopeful that science will help save the reefs, but she is not counting on it. "Ever since I was at university, we have been talking about how in the future, in another generation, the reefs will be gone. But now we're not talking about the future anymore. I will probably in twenty years not see any reefs. I do this work not only for my son, but for me. It's good to have the support of scientists, but we don't have any more time to just sit around and talk. We still don't know which bacteria cause white band disease or why some corals are stronger than others, but we don't have time to wait to figure it all out. We need to just jump in now, and do what we can."

I understand her sense of urgency. After one of my dives, I was gushing to one of the boat's older crew members about the wonders I'd seen below. "Ah," he said mournfully, "you think it is beautiful, but this is really nothing. What you are seeing is maybe ten percent of what once was there." That was sobering.

I admire the work of the Coral Restoration Foundation, as well as the commitment of van Oppen, Gates, and other scientists.* But I'm pessimistic about the future: there is no evidence that the world will address the problems of carbon dioxide emissions quickly enough for coral reefs. If you'd like to see these underwater wonders, you'd better book a scuba trip soon.

*Unfortunately, Dr. Gates passed away in 2018. Her lab colleagues are continuing her pioneering work.

3

A Plague Upon Us

At a sleepover party when I was about twelve, Alfred Hitchcock's *The Birds* came on TV and, sitting cross-legged on the carpet and shoulder-to-shoulder with other flannel-pajamaed girls, I started to watch. I lasted only until the scene — early in the movie, I recall — of screeching seagulls attacking people at a birthday party. I was terrified and snuck upstairs to read. I have never seen the rest of the film.

Hitchcock's story was inspired in part by an incident that occurred in coastal towns around Monterey Bay, California, on August 18, 1961. The *Santa Cruz Sentinel* reported that on that date, residents of Pleasure Point and Capitola were awakened at 3 a.m. by the clattering and thumping of a massive flock of seabirds — sooty shearwaters, to be precise, with three-foot wingspans — diving into the roofs of their homes. The few people who ventured out into their yards to investigate, armed with flashlights, saw large birds littering the streets. Many were dead, but some were only stunned, and those, attracted by the flashlights, flew at the townspeople. In the morning, residents found the streets covered in dead shearwaters and "bits of fish and fish skeletons over the streets and lawns and housetops, leaving an overpowering fishy stench." Thirty years later, a similar event occurred, only that time it was flocks of pelicans — with bellies full of fish — that crashed into houses in the same coastal towns.

In both instances — and no doubt others that went unrecorded — the birds were driven mad by a toxin they'd accumulated in their bodies. The fish they'd eaten had scarfed up zooplankton that had, in turn, swallowed a certain species of microalgae, a diatom called *Pseudo-nitzschia australis.* The diatom produces domoic acid, a powerful neurotoxin that can be lethal to fish, sea lions, and birds that ingest it. The ocean waters had been warm and the winds low both summers — perfect conditions for massive diatom reproduction. It was only a matter of time before their toxins worked their way up the food chain.

Algal blooms that produce domoic acid are increasing in frequency and size globally, as ocean waters warm and fertilizer runoff increases. According to a recent European Commission study, domoic acid poisoning poses a serious threat to marine mammals and humans. Sea lions regularly die of domoic acid poisoning. If we humans inadvertently ingest domoic acid, we suffer vomiting, headaches, disorientation, seizures, and, occasionally, death. In 1987, four people on Prince Edward Island were fatally poisoned and more than one hundred were sickened after eating contaminated seafood. Coastal communities, the study warns, and recreational clam harvesters are especially at risk. Chronic as well as acute exposure puts humans at risk for all these symptoms and for memory problems, too. Because shellfish retain domoic acid as much as a year after exposure to a toxic algal bloom, Washington State advises consumers to limit their razor clam consumption to twelve per month.

Pseudo-nitzschia blooms also cause economic harm. In the summer of 2015, a diatom bloom stretched some 1,500 miles from the central California coast north to British Columbia, producing some of the highest levels of domoic acid ever observed. State authorities closed the region's shellfish fisheries for four and a half months, and consumers were warned against eating any mussels or clams gathered from the shoreline. Fishermen and onshore businesses lost tens of millions of dollars in revenues.

Why cyanobacteria evolved to manufacture toxins is unclear. The poisons weren't originally for defense; cyanobacteria had no predators in their early history. Microbiologists did recently demonstrate that microcystin, the most common cyanobacterial toxin, protects against excessive UV radiation, so perhaps these compounds are sunscreens. In any case, in high doses, they can be lethal, and even a low dose can cause long-term damage to a pet if it makes its way into the animal's liver. There are other cyanobacterial toxins out there, too: Nodularins inhabit brackish water and cylindrospermopsins poison fresh water. While not all cyanobacterial blooms are poisonous, I certainly wouldn't swim in or let my dog drink from any pond that was rimmed with scum or greenly turbid.

And, sorry to say, green is not the only color to be wary of. There's a red-pigmented microalgae, *Karenia brevis,* that goes into overdrive every year along the coast of the Gulf of Mexico and up the Atlantic Coast to North Carolina, tinting the water maroon and bringing physical and economic misery to coastal residents. Because of its hue, blooms of *Karenia* are called red tide, and they are replete with brevetoxins that attack the human nervous system and poison marine mammals. On a weekend vacation to Anna Maria Island off the west coast of Florida a few years ago, I experienced red tide firsthand. Taking a walk along the beach after I arrived, I found myself wheezing, my eyes burning. I wondered if I was sick, but the hotel desk clerk explained—did I imagine with a little chagrin?—that my problem was the local red tide. Although I hadn't noticed a reddish tint to the water, *Karenia* were there in force, vaporized by the action of waves and wind, then inhaled by unsuspecting beachgoers like me. Unfortunately for the communities plagued by them, a red tide can last for months—the 2006 Florida bloom lasted for seventeen months—and shut down the tourist industry. The Florida red tide that started in the fall of 2017 was still causing damage in the fall of 2018, and had killed 2.7 million pounds of fish and other sea animals and sent a record number of people to hospitals for respiratory problems. Because higher concentrations of carbon dioxide in the water (the result of increasing levels of carbon dioxide in the atmosphere) encourage *Karenia* growth, red tide is on the rise.

In recent summers, *Microcystis aeruginosa* and several other toxic cyanobacteria have formed vast blooms in Lake Erie, near Toledo, Ohio. The algae turn the water a nauseating green color and transform its top layer of water into what observers have variously described as pea soup, green paint, or Shamrock Shake. Fertilizer runoff and warmer waters are the primary source of the blooms, but they're helped along by invasive mussels.

Zebra and quagga mussels arrived in the Great Lakes in the ballast water of ships in the 1980s and have been multiplying ever since. They compete with fish for algae, but the mussels are picky, eschewing *Microcystis* while chowing down on its competitors. *Microcystis* therefore has the run of the watery roost and multiplies, releasing its toxins all the while.

The bloom isn't just unsightly; it's a major problem for the well-being of the four hundred thousand residents of Toledo who get their drinking water from the lake. In 2014, Erie recorded its largest bloom ever; it fouled a full three hundred square miles, an area roughly the size of New York City. The intake pipe, which is miles offshore, was covered by the bloom, and the city had to close down its public water system for two days, not only for drinking, but for washing dishes and bathing infants. The city has since installed an advance warning system, so it can start up an additional (and expensive) carbon-activated filtration system, but I imagine Toledans still feel uneasy when they see satellite photos of their water supply completely covered in algae.

The problem is hardly limited to Toledo. Lake Superior and New York's Finger Lakes are now invaded by summer blooms. Utah Lake, a popular recreational lake and water source for livestock near Provo, turned green and had to be closed in July 2016, after the city found concentrations of toxic cyanobacteria over three times the amount considered to be "an acute health risk" by the World Health Organization. That same summer, several people fell sick after swimming in Pyramid Lake outside Los Angeles — health officials measured microcystin levels there at six times the warning threshold. Upper Klamath Lake in Oregon is a poster child for toxic cyanobacterial blooms. Shallow and calm, the lake is an ideal breeding ground for *Aphanizomenon flos-aquae*, whose toxins can cause paralysis. The reservoirs on the Klamath River in Oregon are also prone to *Microcystis* blooms. (Microcystins are particularly dangerous in water used for drinking because they can survive boiling.) The danger isn't just local: the algae pass through the dam's hydroelectric turbines and can be found as far as 180 miles downstream.

Some harmful algal blooms (or HABs) turn water a fluorescent green, or even hot pink, making them easy to spot. But blooms invisibly invade thousands of smaller American lakes, ponds, and reservoirs each year. The US Geological Survey recently found microcystins in 39 percent of streams in the American Southeast, and, in the summer of 2016, nineteen states issued public health warnings because of HABs. The situation in the Northern Hemisphere is getting worse; blooms are developing earlier in the summer and lasting longer. Comprehensive data on the economic impact of freshwater HABs is hard to come by, but a Kansas State University study in 2009 estimated that they cause losses of $2.2 billion annually by limiting recreational water use, devaluing waterfront real estate, and increasing spending to save endangered species and remediate drinking water.

Algae do not kill by toxins alone. Every spring, farmers around the world apply chemical fertilizers to their rapidly growing crops. At the same time, farm animals are born and grow, producing untold tons of manure. Spring and summer rains fall, washing the nitrogen and phosphorus from the fertilizers and manure into streams, which carry them into rivers and then oceans, where they feed algae. In the US, the Mississippi River drains more than a million square miles—40 percent of the American landmass. That means that every year, primarily from March to June, about 1.7 million tons of nitrogen and phosphorus wash off American farms and into the river, and then flow into the Gulf of Mexico. These nutrients create a "super bloom" that permeates a long, roughly rectangular area that stretches from the river's outlet more than three hundred miles west along the Texas coast and twenty miles offshore.

But the super bloom itself isn't the worst problem; it's what happens next, when dead algae become a feast for aerobic bacteria that reproduce with abandon. The worst environmental devastation comes as the bacteria draw down dissolved oxygen until fish, shellfish, zooplankton, and all other oxygen-breathing creatures either swim away or suffocate. Shrimp, for example, die within seconds of encountering hypoxic waters. The algal bloom has, indirectly, created a dead zone, an area of water unable to sustain marine life.

In 2017, the dead zone in the Gulf was the size of Connecticut. Imagine if all the oxygen in the air above the state vanished. It would be empty not only of all birds, but all animals, insects, and other air-breathing creatures on the ground below. This sort of comprehensive destruction happens every summer in the waters off the Gulf, and the devastation is not only dispiriting, it has economic consequences. Fishermen have to travel farther from shore, wasting time and fuel to get to their fishing grounds. And their catch is different: a Duke University study recently demonstrated that when dead zones develop during the summer, shrimpers scoop up more small, lower-value shrimp and fewer large, high-value ones, and fishermen's income declines accordingly.

The same forces — bloom and bust — are at work on the east coast of Florida. Halfway up the Atlantic side of the peninsula lies the Indian River Lagoon system, a priceless and fragile national treasure. The lagoon is 156 miles of shallow water that parallels the Florida coast, flowing slowly between the mainland and a series of barrier islands. Home to at least four thousand species of plants and animals, including a third of the state's manatees and thirty-four other endangered species, the lagoon is also a traditional spawning habitat and nursery for young fish and mollusks. Its economic value has been estimated by the EPA's National Estuary Program at $3.7 billion.

The ecosystem is in grave danger. In 2011, it was hit with two massive algae super blooms. The algae turned the water into thick green goop, blocking sunlight to the bottom and killing 47,000 acres or 60 percent of the lagoon's seagrass, which shelters juvenile fish and shrimp and provides an essential food source for other wildlife. The seagrass loss that year likely cost commercial and recreational fisheries more than $320 million. Similar blooms occurred in 2012 and 2013 and led to the deaths of hundreds of manatees, pelicans, and dolphins. And 2016 brought on the largest and most extensive bloom yet, killing so many fish that their rotting bodies carpeted miles of shoreline. In July 2018, Florida governor Rick Scott declared a state of emergency in seven counties suffering with algae blooms.

The lagoon's problems are exacerbated by the geology of South Florida. The lower half of the state gets fifty to sixty inches of rain each year, and the water table is rarely more than a few inches below ground level, which means the land has little capacity to absorb precipitation. To make matters worse, in many places the underlying limestone is covered by a layer of clay that acts like a sheet of plastic between the topsoil and the rock. The result is that much of South Florida's rainfall runs off horizontally over the ground, into tributaries and stormwater canals. With the water comes farm fertilizers, as well as sewage from six hundred thousand leaking septic systems, systems almost guaranteed to leak, given Florida's geology.

Lake Okeechobee also contributes to the degradation of the lagoon. "Lake O" lies fifty miles west of the lagoon in the center of the state and is almost the size of Rhode Island. Until the early twentieth century, the lake was slowly fed from the north by the waters of the lazy, winding Kissimmee River and smaller tributaries. Overflow and seepage from the lake subsequently resupplied the Everglades to the south.

Over the course of one hundred years, developers and the state of Florida have made dramatic changes to the landscape on all sides of the lake. The marshes north of the lake have been drained and the lazy loops taken out of the Kissimmee to create pastures for dairy farming. The Everglades, south of the lake, have been drained to create sugarcane plantations. Rainwater that once meandered down the Kissimmee and spread out through its surrounding marshes now races into the lake. The lake has been diked to protect the plantations to the south from its frequent overflows, and canals now connect it to the Caloosahatchee and St. Lucie rivers. Today, excess lake water flows west to the Caloosahatchee River and then into the Gulf of Mexico, and east to the St. Lucie River, where it enters the lagoon and the Atlantic.

The water that enters Lake O is polluted with millions of tons of fertilizer, cattle manure, sediment, and sewage. The lake's once sandy bottom is now covered in up to ten feet of mud and sludge, and massive algal blooms are an annual event. Billions of gallons of algae-choked lake water flow into the St. Lucie, especially during the rainy summer months. The river bottom is now also covered deep in muck and its waters are a dead zone. Signs along the river's banks warn people not to fish or swim in it. No one is tempted.

A river is killed; a treasured ecosystem endangered. People focus on the gross algal blooms and blame the algae, but the cause is entirely human. So, what can we do about it?

4

Cleanup

Whether we're talking about Lake Okeechobee, Lake Erie, the Gulf of Mexico, or any other body of water, the best way to avoid algal blooms is to prevent sewage and fertilizers from entering in the first place. But the political will to pass and enforce regulations is in short supply, and the immediate costs and inconveniences obscure the longer-term benefits. Politicians in Florida, for example, have been promising to reduce the flow of phosphorus into Lake Okeechobee for more than thirty years. Finally, in 2005, the legislature passed a law requiring that the mineral load be reduced from an average of 500 tons to 140 tons per year by 2015. But at the deadline, the load was unchanged. The legislators' response? With the support of farm lobbyists, they simply moved the date of compliance forward ten years.*

As an alternative, some ecologically minded entrepreneurs are trying a new approach: intercepting the excess nutrients after they've entered streams, rivers, and canals, but before they reach the lagoon. When I learn they are enlisting algae to do the job, I have to check that out. That's why I'm driving on a July morning to Vero Beach, a city that spans the Indian River Lagoon, about halfway down Florida's Atlantic coast. I am going to meet Allen Stew-

---

*Construction of a 10,500-acre reservoir south of Lake Okeechobee, on land that the state would purchase from US Sugar, Florida Crystals, and other sugar companies, is awaiting funding. The goal is to capture lake overflow that currently goes east to the Indian River Lagoon and west to the Gulf, treat it, and send clean water south to the Everglades where—because the Lake O dike stops southward water flow—it is desperately needed. Of course, it would be most helpful if farmers and municipalities north of the lake would rein in pollutant discharge. In any case, the accumulated phosphorus at the bottom of Lake O will continue to leach out for many decades.

art, one of the founders of HydroMentia, Inc., an innovative water pollution control company.

HydroMentia's origins lie in the research of Dr. Walter Adey, director of the Marine Systems Laboratory at the Smithsonian Institution in Washington, DC. In the 1970s, Adey was studying the ecology of ocean reef systems, fascinated by the way they recycle nutrients. To fully analyze and understand the processes, he began creating enclosed reef microcosms, starting with tabletop aquariums and ultimately building the forerunners of magnificent artificial reefs like the two-million-gallon tank at the National Aquarium in Baltimore.

All aquarists, whether they're caring for a goldfish bowl or a three-story reef replica, must deal with the lethal buildup of ammonia, nitrates, and phosphates from their animals' excreta. Hobbyists and professionals traditionally either changed the water in their tanks frequently or treated it with chemicals. Adey recognized that in the wild, turf algae help keep bodies of water in chemical balance by capturing those nutrients. Marine animals then graze the algae, and thus slowly recycle the minerals.

Adey figured out how to replicate nature on a pint-size scale for aquariums, and invented and patented a device he called an Algal Turf Scrubber. His original ATS was a plastic box primed with a little turf algae and attached to the outside of a fish tank. Water circulated from the aquarium via plastic tubing into and through the ATS; algae absorbed the nitrogen and phosphorus and grew, all the while sending clean water back to the aquarium. When the algae completely filled the box, the aquarist scraped them off, and started over.

In the early 1990s Adey recognized that the principles behind his ATS might have a broader application, cleaning larger, partially enclosed bodies of water damaged by excess nutrients. He was inspired in particular by the Chesapeake Bay, which was close to his home and work and was badly polluted by runoff. In 1996, he, engineer and biologist Allen Stewart, and several investors founded HydroMentia to adapt and apply Adey's technology on a vastly larger scale.

• • •

Allen Stewart is driving across Florida from his current home in Punta Gorda to give me a tour of a HydroMentia ATS. We had agreed to connect in the parking lot of a Cracker Barrel and ride together to the site, and he had emailed me a photo of himself, so I could identify him. "I'm old and ugly, as you can see," he'd written, but when I looked at the image of a man seated in a kayak, his appearance was not what caught my eye. It was the spotted fish he was holding in his outspread arms, a fish so big it sagged between his right hand, which was grasping its tail, and his left, which held its head. The photo is telling: Allen Stewart is in the business of water cleanup because he has always been a passionate outdoorsman and has a long-standing love of Florida's wild places.

At the appointed hour, we find each other and drive a few minutes to Osprey Marsh, just inland of the lagoon where an Algal Turf Scrubber is installed. I've been imagining a major piece of machinery, a giant version of Adey's aquarium model, but when we pull up to the ATS, all I see is what looks like a wet, concrete parking lot. We walk to the edge. Its four-acre surface is covered with a sheet of shining water that mirrors the deep blue sky and the herd of popcorn clouds above. The technical term for this expanse of concrete is a floway; it's ever so slightly slanted from one end to the other so the water arriving from land at the top end moves very slowly and evenly down its length. Stewart tells me it takes fifteen minutes for water to make the journey.

The floway is, in essence, an exceedingly shallow lagoon. At this moment, a dozen white egrets step carefully across its surface, dipping their yellow beaks into the water to snag a tiny fish or crustacean. A flock of sanderlings — small gray shorebirds — wheels in the sky and then settles down en masse. Stewart points out two little blue herons, two great egrets with black legs and orange beaks, a few glossy ibises, and a coterie of American coots — black, duck-like birds with a vertical blaze of white on their steep foreheads. In winter, he tells me, I might see flocks of pink-feathered roseate spoonbills.

We're not here for the birds, the ducks, or the little fish, though. We're here for the algae. As I crouch at the edge of the floway and peer below the surface, I see that little filaments of the stuff carpet the concrete in a fur of green, brown, and gold. Stewart explains that most turf algae are green species, but that brown diatoms often cling to them, obscuring their color. At Stewart's suggestion, I reach in and grab a clump to inspect. It's cold and mushy, but full of hidden treasures for hungry birds: tiny clams no bigger than a nail head and amphipods – shrimplike creatures the size and shape of a fingernail paring. If I were to look at a bunch of turf algae under a magnifying glass or microscope, Stewart tells me, I would see that it has a matted texture because the filaments are stuck together by the mucilage on their surfaces. The floway is an ideal habitat for these algae: The sunlight is bright, the water too shallow for any marsh plants to grow and shade them, and the descending current just fast enough to prevent suspended microalgal blooms that might block the sun. Best of all, the water, supercharged with nitrogen and phosphorus from terrestrial sources, is algae elixir.

We walk to the higher end of the floway, where the water enters. The open mouths of dozens of white pipes gape along the top edge of the concrete. Every minute or so water surges out of all of them simultaneously, pushed by a small pumping station, and a new sheet of water skims slowly toward the other end.

Two sources feed the pipes. The South County Water Treatment Facility next door sends 1.5 million gallons a day of "reject water" to the floway. Reject water has been cleaned of its organic compounds and is perfectly potable, but it still contains high levels of nitrogen and phosphorus. The other source, which sends more than 10 million gallons per day toward the lagoon, is the South Relief Canal, one small piece of the system of thousands of miles of freshwater canals, levees, berms, and pumping stations that crisscross the southern half of the state, collecting and routing stormwater to the coasts and away from farms and neighborhoods.

I have arrived on a harvesting day, and as Stewart and I watch, a small green tractor with a blade attached to its front "mows" the eastern half of the floway. It trundles noisily from the edge toward a central channel, pushing water and algae with it, then backs to the edge to push again. Mowing an ATS is more like plowing slush than cutting grass.

We walk about five hundred feet to the low end of the floway, where the algae slush slips from the channel onto a broad, perforated conveyor belt. The belt tilts upward, and water gushes from its underside as the algae drain. At the top end of the belt, the algae drop about twenty feet to a concrete pad. There, a fellow with a rake spreads the biomass over the concrete to dry.

That's all there is to an ATS system; it couldn't be lower tech. Simple as it is, Stewart tells me that "over the course of 2016, this floway will produce almost seven million pounds of wet algae. It's only our first year at Osprey Marsh, but based on the data so far, that means we'll be taking about five thousand pounds of phosphorus and fifteen thousand pounds of nitrogen out of the water. That's twenty thousand pounds of nutrients that never enter the lagoon."

All in all, as a demonstration of the potential of algal turf scrubbing technology, the Osprey ATS and its sister ATS at nearby Egret Marsh are impressive. However, to put that accomplishment in perspective, while the scrubbers clean about seven billion gallons of water annually, the total flow from Lake O through the St. Lucie to the lagoon is about 250 billion gallons. It will take many more turf scrubbers to make a significant dent in the lagoon's problems.

Trapping pollutants before they get into open water is not a new idea. Nature invented wetlands — marshes, swamps, and bogs with their lush aquatic plants — that have been absorbing nitrogen, phosphorus, and other nutrient runoff for millions of years. But Florida has lost nearly 50 percent — more than nine million acres — of its natural wetlands, and therefore half of its ability to trap and recycle nitrogen and phosphorus. And, of course, wetlands have far more nutrients to trap today than they did before the state became so densely populated and intensively farmed. Recognizing this unnatural disaster, in the 1990s the state legislature called for the creation of "constructed wetlands" as a partial remedy for nutrient overload. Constructed wetlands are areas — sometimes sunken and filled with gravel and sand — enclosed by levees that naturally fill with marsh plants that take up nutrients.

There are now fifty-seven thousand acres of constructed wetlands along south Florida's estuaries and bays.

But while constructed wetlands are useful for sequestering nutrients, and valued for the wildlife they attract, Stewart points out that they are less than ideal. "Really all we're doing is changing the location of the phosphorus. It goes from farms to water to aquatic plants in various kinds of constructed wetlands. The plants grow, die, and decay, and their remains settle to the bottom sediment of these wetlands. So, the phosphorus just gets transferred." (Phosphorus is the worst problem here. In mucky, anaerobic wetlands, if fixed nitrogen isn't absorbed by plants and algae, denitrifying bacteria convert it back into harmless nitrogen gas that returns to the air.) When tropical storms hit the area, they churn the accumulated sediments, releasing phosphorus into the water. Even without storms, about 25 percent of the sequestered metal slowly leaches out into the water column. Many of the artificial wetlands already have so much phosphorus that cattails, which thrive in the enriched and ever-deepening muck, are taking over. They now have to be sprayed with herbicides, otherwise their dense thickets crowd out the natural sawgrass, block light from reaching the water, and change the natural habitat critical for native wildlife.

The manmade wetlands will eventually—the time frame is several decades—fill up entirely with sediments. People talk vaguely of dredging them when the time comes, Stewart says, but the task would be gargantuan. And where would they put all the muck? Stewart's opinion on constructed wetlands: "It's like a man falling off the top of a building. On the way down, he passes people standing on their balconies who ask him how things are going. He keeps answering, 'So far, so good!' The problem is, no one is taking the long view here." Stewart makes the case that algal turf scrubbing is, at the least, a complementary technology to the artificial wetlands. The scrubbers remove ten to forty times more pollutants per square meter than artificial wetlands do, and they remove them permanently.

The facilities at Osprey and Egret marshes are the only two full-scale ATS systems in operation. Stewart and his colleagues are giving away the technology in the hopes of encouraging more scrubbers along the coast at canal outflows and next to water-treatment facilities. But Stewart has even bigger dreams: He would like to see a ring of scrubbers circling Lake Okeechobee. Not only would they intercept water coming from the north via the Kissimmee River and other tributaries, they could, with small pumping systems, pull water from the lake, clean it, and send it back. "They would work like a body's kidneys, constantly removing pollutants." He knows, though, that such a grand scheme won't be happening anytime soon. The up-front costs of engineering and construction, he acknowledges, as well as the purchase of lakeside land, would be issues. Nonetheless, he says the lack of political will is an even greater obstacle.

But consider the alternative. In February of 2016, after an unusually wet winter, Lake Okeechobee was in imminent danger of overflowing to the south and inundating the region's sugarcane plantations. The Army Corps of Engineers made the decision to throw open the floodgates from the lake to the St. Lucie. The rate of discharge rose from four hundred million gallons a day to more than one billion gallons. Along with the water went nutrients and algae from the lake's two-hundred-square-mile bloom. By midsummer, large parts of the river, the lagoon, and even ocean beaches east of the lagoon were clogged with toxic, decaying algae thick as guacamole. Fish and manatees died, the stench was nauseating, and tourists stayed home. Estimated losses were $470 million. There is a cost to building ATSs, but we pay a high price — and the price will only go up — in doing nothing.

Dr. Patrick Kangas, associate professor in the Department of Environmental Science and Technology at the University of Maryland, has been running temporary pilot ATS systems all around the Chesapeake Bay for ten years. The bay is polluted with nutrient runoff from agriculture, poultry farms, and the suburban lawns in its sixty-four-thousand-square-mile watershed. In the middle of summer, the bay's dead zones represent about 10 percent of its total volume. Despite three decades of efforts to improve the bay's water quality, little progress has been made.

Kangas, a graying bear of a man, is committed to using engineering to help solve ecological problems. The day I stop by his office, he is upbeat. "The EPA just certified algae turf scrubbing as a 'best management practice' for controlling pollution in the bay. I've been harvesting algae every week of the growing season at pilot projects around the bay for ten years. Me," he emphasizes, "me personally, I've been harvesting, and I'm not getting any younger. So the fact that EPA certified it is a *really* big deal."

In Maryland, certification means that municipalities can use ATS systems to meet the EPA's established limits on the amounts of fixed nitrogen, phosphorus, and sediments that a city or town can discharge into the bay. Kangas and others are now working with the Port of Baltimore on setting up a full-scale ATS in a five-hundred-acre paved area where ships unload vehicles and park them before they're trucked to dealers. That area is also a major conduit for stormwater runoff. One of Kangas's pilot studies revealed that because an ATS is so efficient, it will take a mere half-acre ATS floway to clean the stormwater.

"One of the beauties of the ATS system," Kangas says, "is that it's fundamentally simple, with free energy from the sun and free algae from the very water that is being cleaned. It's great because you're putting just a little energy into a natural system, and it does these cleanup services for you. This is why ATS is so much cheaper than traditional wastewater cleaning systems. We're also investigating putting the algal biomass into an anaerobic digester [a simple device in which bacteria break down organic compounds] to make methane gas to run the pumps."

Which is not to say that the engineering is a no-brainer. "There are important engineering decisions involved in each situation, and we're nowhere near optimizing the basic system. We just haven't had enough money to get there yet. But now, regulation is making a market for stopping these nutrients before they hit the water. Finally, there's a financial incentive to perfect and use ATS."

• • •

So, at the end of the day, what to make of Algal Turf Scrubbers?

For one, they work: Scientists at the US Department of Agriculture have found that they can capture 60 to 90 percent of nitrogen runoff and 70 to 100 percent of phosphorus runoff. ATSs require little energy, have sturdy components, and incidentally provide a habitat for wildlife. Because all algae are welcome on the floway, ATS operators face none of the problems algae farmers do; as the seasons change, the crop spontaneously adapts. Because the algae are filamentous and stick together, they're easy to harvest, and they dry in the sun. Removing pollutants from single-point sources (like a factory) is an easy task compared to cleaning water from many diffuse sources (like farm runoff or leaking septic systems). Turf Scrubbers are uniquely able to capture nitrogen and phosphorus from nonpoint sources before they hit open waters.

But what would we do with all that harvested algae? The biomass performs as well as commercial fertilizers, according to the USDA, and a small quantity from Egret Marsh is currently sold as a soil supplement and a potting soil. Twenty years ago – the last date for which figures are available – a survey found that Florida businesses and farms used about forty-two million cubic yards of compost in landscaping, packaged soils, nurseries, sod and tree farms, and agriculture per year. At the time, Florida-produced compost met less than 20 percent of that need. So, it's safe to say there would be demand for more of Florida's ATS-generated compost right at home.

Biofuel and plastic manufacturers might be able to use biomass from scrubbers, but only with further treatment; turf algae usually trap too much sediment for these buyers. But Algal Turf Scrubbers might be engineered to produce better, more marketable biomass. Professor David Blersch, a biosystems engineering expert at Auburn University, explained to me that he, Dr. Dean Calahan at the Smithsonian, and others are working on that possibility. They are focusing on altering the floway surfaces – now poured concrete – to improve the quantity and quality of the turf algae that grow on them. They've been testing three-dimensional plastic grids that would grow turf algae more densely, so floways could have a smaller footprint. Blersch is also investigating whether, using 3D printers, he can create a raised pattern (mere microns in elevation) on a plastic substrate that would selectively attract algae species — say, diatoms naturally rich in lipids. "My hope," Blersch says, "is that with this material we could grow 80 percent of a particular species, and could give that to a company to make plastics or biofuels, depending on its characteristics."

While I hope the engineers are successful and ATS biomass has a market value, I doubt that the scrubbers will make a profit anytime soon. For one, they usually require lots of land, often in places where acreage is expensive, like the shores of the Chesapeake Bay. So, we will need regulations and taxes to provide the incentives for building and operating them. That will take citizens — appalled by nauseating algal blooms, the stench of dead fish, the decline of property values, and the loss of tourist revenues — demanding that local governments create those incentives.

5

Making Monsters

Algae are powerful enough on their own to change life on the planet. Billions of years ago, they oxygenated the oceans and the air, and then sent the planet into a deep freeze. They killed or banished the oceans' anaerobic organisms and covered the land in plants. But they did their work slowly, over eons. Today, with mankind in the mix, algae are operating at warp speed.

Fly high over the Baltic Sea in midsummer, and you will see the world's largest algal bloom, covering about 145,000 square miles, or the size of the state of Montana. From far above, its surface looks beautiful, all green whorls and swooshes like the marbleized endpapers of an antique book. At ground level, the reality is that the Baltic Sea is pocked with dead zones, and its fisheries are collapsing. Researchers are tracking this and four hundred other significant dead zones across the world, and the forecast — with 100 percent likelihood — is for more and bigger blooms ahead, all thanks to us.

Recently, algae have erupted in a new and insidious manner, hidden inside an explosion of *Noctiluca scintillans,* a heterotrophic dinoflagellate. *Noctiluca* are transparent, spherical, and tiny, only a millimeter in diameter. Not photosynthetic themselves, they engulf microalgae and take them into their cytoplasm where they live as symbionts. The algae pass sun-derived sugars to their zooplanktonic hosts in exchange for the nitrogen compounds and carbon dioxide they generate from their food. In the past, *Noctiluca* went unnoticed by the world, except for sailors who occasionally spotted their spectral bluish bioluminescence at night. But today, these zooplankton are despoiling miles and miles of ocean in the Arabian Sea with green scum.

Why now? According to a report from the Center for Climate and Life at Columbia University, rising carbon dioxide levels in the atmosphere have triggered a cascade of ecosystem changes that have been sending *Noctiluca* populations soaring. The problem started as the atmosphere above the Indian subcontinent grew hotter, increasing the difference between land and sea temperatures. That differential increases the strength of the monsoon winds blowing toward land, and the higher winds increase upwelling, which brings more nutrients to the surface. Algae then go into overdrive, which means that decomposing bacteria and other heterotrophs are also flourishing.

Noctiluca do especially well in the new conditions. For one, they can turn to solar power when other heterotrophs, having eaten all the algae on offer, starve. Moreover, as they prosper and spread, their blooms shade out diatoms and other phytoplankton, starving the zooplankton that dine on them. Even better (only for the *Noctiluca*), they have a secret defensive weapon: They produce high levels of ammonia, which irritate the digestive tracts of fish. The fish, no fools, consequently avoid the blooms.

The *Noctiluca* are a triple-barreled peril for mankind. Fishermen pull up empty nets because the lowest levels of the marine food chain are disrupted and the larger fish have fled in search of food. Tourists flee, as well, to beaches unmarred by scum. And, in the Arabian Sea, the scum clogs the intake pipes of critical desalination facilities. In January 2018, a *Noctiluca* bloom in the Arabian Sea covered an area three times the size of Texas. Similar blooms have become a problem in the warm waters off India, Southeast Asia, Africa, and Australia.

That global warming is the result of burning fossil fuels is indisputable, but there's a great deal we still don't understand about how the changing climate will affect conditions in particular regions of the world. The Greenland ice sheet has been shrinking for a hundred years and the annual loss has doubled in the last fifteen years, but geoscientists have not determined exactly why the Arctic is warming faster than temperate and tropical zones. Arctic climate change models have included, of course, the fact that as ice and snow fields retreat, they expose more dark-colored land and open ocean, both of which absorb far more light — and therefore retain more heat — than pristine snow. But even when climate change modelers incorporate all relevant factors, it is clear that other factors influence the rate of temperature change in the highest latitudes.

The scientists at the Helmholtz Center Potsdam–German Research Centre for Geosciences recently reported that one of the missing variables in the equation is — you guessed it — algae.

It's to be expected that some microalgae can survive freezing conditions; after all, all algae descend from survivors of planetwide glaciations. One cold-adapted species, *Chlamydomonas nivalis* or snow algae, goes dormant in the winter, but as soon as sunlight heats the surface snow just enough to create a little meltwater, it springs to life. And when it does, it blooms in pink.

Why pink? It's the color of astaxanthin, a carotenoid pigment that protects *nivalis* from the flood of UV light that strikes the high latitudes in the summer. Should you come across an area of rosy snow, you will see it is noticeably sunken: the algae absorb solar energy, heating themselves and the snow around them. "Watermelon snow" delights mountain hikers around the globe, but it reduces snow's reflectivity by 13 percent.

Scientists have just begun to fully appreciate the impact of watermelon snow on climate change. A team of researchers recently reported in *Nature Geoscience* that *nivalis* accounts for about 17 percent of the annual melt on Alaska's 750-square-mile Harding ice field. Algae account for 5 to 10 percent of the Greenland ice sheet dissolution. As the climate warms and meltwater appears earlier in the spring and lasts later in the fall, algae will have a longer growing season, leading to even more ice melt, spurring a feedback loop that hastens the disappearance of ice fields and snow-covered areas.

While *nivalis* is changing the Arctic, the diatom *Didymosphenia geminata,* commonly known as didymo, is invading the underwater landscape of some of Earth's most pristine temperate rivers and streams. However destructive watermelon snow is, at least it charms the eye. That cannot be said of didymo, whose well-deserved epithet is "rock snot." First discovered in rivers on Vancouver Island, British Columbia, in the late 1980s, the yellowish-brownish mats now blanket streambeds and riverbeds around the globe. Over the last two decades, like creatures from a Roger Corman horror movie, rock snot has overrun tributaries in countries as far-flung as Chile, the US, Canada, Iceland, Poland, and New Zealand.

Under normal circumstances, an individual diatom extrudes a single sticky, threadlike stalk to attach itself to a rock or plant. But sometimes after settling down, diatoms then send out a mass of additional stalks that entangle with their neighbors'. A grotesque invasion of rock snot ensues.

As didymo outbreaks multiplied at the turn of the twenty-first century, scientists suspected that fishermen might be spreading didymo when they trekked from one stream or river to another. It seemed that the felt boots many wore might be a particular hazard. So, in the US, around 2010, state recreation departments began posting signs along affected watercourses, urging fishermen to clean and dry their canoes and gear and discard their felt boots in order to contain the epidemic. Maryland, Vermont, and other states banned the boots entirely. Nonetheless, rock snot continued to extend its territory.

Which is not surprising. New research indicates that fishermen are not to blame for the plague. (Vermont repealed its ban on felt soles in 2016.) Recently discovered fossils prove that the diatoms have inhabited many of the affected waters for hundreds and even thousands of years. Instead, changing environmental conditions have sparked their population explosion. Scientists now believe that lower phosphorus levels in these pure streams provoke the outbreaks.

Why is there less phosphorus in these waters? Once again, we start with humans burning fossil fuels. This time the damage comes when, in the heat of combustion, nitrogen and oxygen in the air react to form various nitrogen oxides (or NOx), components of smog and a lung irritant. Some NOx compounds undergo further chemical transformation and settle out of the air and onto earth.

In forests, that means soil organisms have access to more fixed nitrogen, which they readily take up and consequently use to reproduce prolifically. As they do, they also take up more phosphorus, which means there's less of the mineral in the soil to wash off the land and into streams. Didymo, desperate for phosphorus, sends out more sticky stalks in an attempt to capture what they need. The results are disgusting.

It's no wonder that the first didymo outbreak appeared in British Columbia. The provincial government has been fertilizing hundreds of thousands of acres of pine forest by dropping nitrogen pellets on them from helicopters. Why fertilize the forests? The lodgepole pines are a significant source of income for the timber industry, but they've been decimated in recent years by infestations of mountain pine beetles. (The beetles are surviving the winters in greater numbers, thanks to the rise in global temperatures.) By adding nitrogen to the forest, the remaining trees are spurred to produce more wood. But, the pines are also depleting phosphorus reserves from the soil, which means less ends up in streams.

Climate change likely plays another role in the surge of didymo in northern estuaries: as temperatures rise, less ice forms, which means more sunlight reaches the microalgae, providing them with more energy to multiply. And, in rivers that have traditionally been hatcheries to wild salmon, the decline of the fish contributes to the didymo problem; with fewer adults returning to their freshwater birthplaces to spawn, die, and decompose, less phosphorus returns to the water.*

It's not clear how harmful didymo is. Ecologists worry that the algae diminish trout populations by preventing them from success-

---

*New Zealand's South Island is one place where fishermen may well have contributed to a didymo invasion. The country's North Island is volcanic, so its rivers are rich in phosphorus and therefore are not troubled by didymo. The South Island, on the other hand, is not volcanic and its waters are both clean and low in phosphorus. Didymo is not native to the island, but a decade ago fishermen and tourists started traveling there in large numbers to explore the wilds made famous by the *Lord of the Rings* movies. It appears the visitors brought didymo with them and likely helped spread it.

fully attaching their egg sacs to underwater rocks. In a similar way, they may also reduce the numbers of caddisflies and mayflies that the fish dine on. The jury is still out, but the didymo plague is a reminder that no place on the planet, no matter how remote, is immune to the effects of climate change.

The spread of toxic microalgae is also having an impact in unexpected places and unforeseen ways. The EPA notes that climate-induced drought in the American Southwest is raising the salt concentration in some freshwater ecosystems, allowing toxic marine algae to grow in lakes. And the reverse is true, too: Toxins from HABs in freshwater lakes are being washed into offshore waters. In 2013, researchers discovered that freshwater algae toxins are accumulating in marine mussels in Puget Sound. In Monterey Bay, California, sea otters are regularly found dead, poisoned by freshwater *Microcystis*.

Some seaweeds have gone into overdrive, too, nurtured by warmer and more nutrient-rich offshore ocean waters. Massive aggregations of free-floating *Sargassum* (called golden tides), which began washing ashore on the beaches of the Gulf of Mexico, Bermuda, and the Caribbean Islands in the 1970s, are growing larger and arriving more frequently. In 2015, ten thousand tons *per day* arrived on Caribbean beaches and made their first appearance in Brazil and on the northwest African coast.

Shoals of bright green *Ulva* now periodically blanket beaches around the world; in 2008, international TV viewers watching the 2008 Olympics in Qingdao, China, were amazed to see how more than a million tons of the seaweed covered the Yellow Sea. Only the efforts of some twenty thousand Chinese volunteers and an armada of small boats managed to clear the waters for the sailing events. The media focus has moved on, but the "green tides" are still a major problem.

Not only is it expensive for communities to truck away heavy drifts of seaweed, tourists naturally avoid beaches covered in mounds of the stuff that quickly decompose in tropical heat. There's a potential health hazard, too: anaerobic bacteria in the interior of the biomass release toxic hydrogen sulfide gas.

In the future, inevitable increases in nitrogen and phosphorus will make all algal blooms worse. The human population is growing, which means we are farming more marginal land and fertilizing crops more heavily. Because a warming atmosphere holds more moisture and therefore produces heavier rains, ever greater amounts of these nutrients will be sluicing into tributaries, lakes, and oceans. Nitrogen levels in American rivers and other waterways are projected to increase by 20 percent on average by the end of the century, with larger increases in the Mississippi River basin and the Northeast.

Spurred by greater energy in the warming atmosphere, average wind speeds off coasts – including America's West Coast – have been rising in recent years. The resulting upwelling brings more nutrients to the surface, feeding blooms. When the blooms cover the water, they absorb sunlight, further heating the surface, which encourages the growth of even more algae. Another harmful side effect: the warmer waters act like a lid on the ocean, preventing atmospheric oxygen from penetrating to deeper waters, making them less hospitable to fish and other aquatic animals. As seawater levels continue to rise in the coming decades, they will create ever larger shallow-water expanses off our coasts – perfect habitats for more algal blooms, toxic and otherwise.

The fact is we're turning mild-mannered algae into ever more deadly monsters.

6

Geoengineering

In 2015, 196 countries signed the Paris Agreement, pledging to take steps to reduce the growth of carbon emissions in order to limit the global average temperature rise in the current century to 1.5 degrees Celsius (2.7 degrees Fahrenheit). Nations were to meet their individual goals by whatever means — regulation, carbon taxes, or cap-and-trade regimes — they deemed best. But, judging by progress to date, there is almost no chance we will meet that goal. In fact, according to a UN Intergovernmental Panel on Climate Change (IPCC) report issued in October 2018, we will reach that 2.7 degree Fahrenheit mark by 2040. By the end of the century, average global temperature will have risen as much as 4 degrees Celsius (7.2 degrees Fahrenheit).

Already, people around the world are experiencing the bitter effects of climate change. Jakarta is submerging and Pacific Islands are disappearing. Residents of Miami regularly find their streets flooded when the moon is full, and in Norfolk, Virginia, at least once a month, seawater bubbles up through storm drains to pool in roads. In coastal Alaska, villages once protected from coastal storms by miles of sea ice have already had to move inland. Two recent studies attribute about 20 percent of the rain that fell on Houston during Hurricane Harvey in 2017 to climate change, and initial data indicate that Hurricane Florence was about 50 percent wetter than it would have been in the absence of global warming. The oceans will continue to rise: the IPCC uses five models that project a range from 0.4 meter to 1 meter (1.3 feet to 3 feet). National Oceanic and Atmospheric Administration (NOAA) scientists concluded in 2012 that there is a greater than 90 percent chance that global sea level will rise as much as 6.6 feet (about 2 meters). Worldwide, the number of people displaced from coastal areas will certainly number in the hundreds of millions.

At the same time, the American West is suffering under an extended drought; rampant forest fires are regular and deadly summer events. The Swedish forest fires of 2018 — some burning north of the Arctic Circle — are harbingers; polar regions are warming twice as fast as the rest of the planet, and such conflagrations will become ever more common. The 2003 heat wave in Europe killed at least 35,000 people, and a recent study published in *Nature Climate Change* predicts that, even with emissions reductions, 75 percent of mankind will face at least twenty days a year of similar temperatures by the year 2100.

We'll be getting sicker, too. Global warming spurs Lyme disease by speeding the maturation of disease-carrying ticks; their reproduction rate has nearly doubled in the US and quintupled in Canada. Tropical diseases — West Nile, Zika, and malaria, for example — are spreading northward as mosquitoes survive milder winters. *Vibrio,* a warm-water bacterium, is expanding its range, too. According to the US Centers for Disease Control, every year the bacterium infects an average of 80,000 Americans — and kills 100 — who eat undercooked seafood or are exposed to seawater through a break in the skin. *Vibrio* infections are now striking people who live as far north as Alaska, Sweden, and Finland. Cases of leishmaniasis, a tropical disease that causes liver damage, are showing up in northern Texas. It's certain that, fifty years from now, every person on the planet will be affected — for the most part adversely and in many cases severely — by climate change.

So, it takes no great leap to imagine that by 2100, governments will be willing to take drastic action, not just to limit new carbon dioxide emissions, but to remove billions of tons of the gas already in the atmosphere. Scientists are already beginning to investigate large-scale "geoengineering" options. Some experts, like Klaus Lackner, propose mechanically removing CO_2 from the atmosphere. Others are experimenting with chemically capturing the gas in Iceland's volcanic basalt and Oman's peridotite rock formations, speeding up a natural but slow mineralization process. Still others believe that we can use algae to counter climate change; after all, algae have been helping cool the planet by burying carbon on the ocean bottom since the day the first cyanobacterium appeared. Occasionally — as during the global glaciation 2.4 billion years ago — they do so in radical fashion.

More benignly, 49 million years ago, algae cured a hothouse Earth and left us the story of how they did it, written on the seafloor. At the time, carbon dioxide levels were a hundred times greater than they are today, and the planet was entirely ice-free. Tectonic plate movements had carried most of Earth's landmasses northward, clustering them near the top of the globe, where they almost completely enclosed the Arctic Ocean and turned it into an inland sea. Even at the continents' northern shores, hippos trundled about and crocodiles and aquatic snakes patrolled the waters. In the junglelike atmosphere, rain fell frequently and heavily, and freshwater rivers gushed into the warm ocean, washing nutrients off the land. Most surprisingly of all, in summer the Arctic Ocean — some 1.5 million acres — was covered shore to shore in a thick, vivid green carpet of azolla, that tiny fern and its symbiotic cyanobacteria that I grew in my pond in Maryland.

Azolla is a freshwater plant, so how could it thrive in ocean water? The answer lies in physics: Because fresh water is less dense than saltwater, the incoming river water tended to float on top of the seawater. Ordinarily, wind generates enough turbulence to mix such layers, but at the time, the blanket of azolla calmed the surface, so the freshwater layer stayed intact. At the end of each summer, as the long daylight hours dwindled, the entire mass of ferns died and sank. Because the sea below the surface was anoxic, there were few zooplankton and bacteria to feast on the ferns' remains, and most of the biomass fell to the seafloor. Each spring, a new crop of azolla took over the surface.

This Azolla Event, as it is called, persisted for 800,000 years. All those annual blooms were buried in sediment and compacted over the ensuing millions of years, ultimately becoming the billions of barrels of crude oil that lie under Alaska today. But of far greater consequence, these tiny plant/algae hybrids sucked in and sequestered carbon dioxide, reducing the concentration of the gas in the atmosphere by about 80 percent in less than a million years—a notably speedy remediation. The Azolla Event brought about the return of the planet's ice caps.

In the last million years, our planet has experienced ice ages every hundred thousand years, and it appears that algae are responsible. In 2016, researchers at Cardiff University were examining layers of fossilized marine algae and found that some had stored a much higher level of carbon than others. Those carbon-rich layers occurred every hundred thousand years. It looks like algae periodically remove so much carbon dioxide from the atmosphere that large ice sheets form in the Northern Hemisphere. In other words, algae do a regular spring cleaning of the atmosphere.

So, the question is: Since algae are so effective at cooling our climate, can we enlist them to do so now?

Algae are not equally distributed throughout the world's oceans, and the Southern Ocean is notably lacking in them. Dr. John Martin, director of the Moss Landing Marine Laboratories in California, discovered in the 1980s that algae are scarcer there because the ocean lacks iron, a metal critical to photosynthesis and other cell functions. And it's not just the Southern Ocean; nearly one-third of the world's oceans have some degree of iron deficiency.

Martin once said, "Give me a half tanker of iron, and I will give you an ice age." A joke, yes, but what he meant was that if we add iron to iron-deficient waters, greater numbers of algae will flourish and then will sequester carbon dioxide on the ocean floor. Add enough iron to the Southern Ocean and, theoretically, algae will increase thirtyfold. That would—again, theoretically—mean a thirtyfold increase in the amount of carbon dioxide the Southern Ocean could remove from the atmosphere. Core samples taken from ancient ocean beds indicate that, in the past, higher levels of marine iron, greater numbers of marine life, and lower levels of carbon dioxide occurred together.

Although iron molecules tend to sink, ocean waters are constantly "fertilized" with new iron, raised from the depths by upwelling, swept in from eroding land, or deposited by volcanic eruptions. In November 1991, Mount Pinatubo in the Philippines erupted dramatically, providing scientists with an opportunity to measure a natural experiment in iron fertilization. Over the course of about three months, the volcano shot a plume of ash containing more than forty thousand tons of iron into the upper atmosphere. Most of that iron settled into the Southern Ocean, where it indeed sparked a massive, long-lasting algal bloom. The bloom sent a measurable pulse of oxygen into the atmosphere and reduced carbon dioxide.

John Gribbin, a British astrophysicist, was the first to suggest, in a letter published in *Nature* in 1988, that "by adding iron compounds to the ocean, a 'technological fix' to remove carbon dioxide from the atmosphere might be possible." The idea makes sense, but technical questions abound. How big should the iron particles be to keep them suspended in the water long enough to be incorporated by algae? How much iron dust would it take? Equations predict how much CO_2 the algae could take up, but how much would actually be sequestered? Would the additional algae simply be eaten by zooplankton that then quickly release carbon dioxide through cellular respiration? And how would you measure the results of any studies done in the open ocean?

To find the answers, Martin's colleagues at Moss Landing undertook the first iron-seeding experiment off the Galápagos Islands in 1993. (Martin helped prepare for the experiment, but, sadly, died before the ship sailed.) Since then, scientists — plus an American businessman who hoped to revive a Pacific salmon fishery — have run thirteen open-ocean iron fertilization experiments. The method they've developed involves spreading iron dust off the back of a ship into an ocean eddy, which corrals the iron so the impact on algae populations can be studied.

In fact, blooms develop and can last for months, but the most important information — how much carbon the algae take to the bottom — is still uncertain. Some of the ambiguity stems from the difficulties in assessing results. Scientists use turbidity meters that measure the loss of water transparency due to suspended particles — such as dead algae. But, algae sink slowly; it can take months for their carbon remains to reach the ocean floor, by which time currents may have carried them far from the iron-seeding site. Certainly, zooplankton eat a portion of the new algae, but even then, some unknown percentage of the algal carbon still makes it to the ocean floor, either in fecal pellets or as zooplanktonic remains.

Despite these difficulties, the experiments have revealed much useful information. For one, it takes a relatively small amount of iron to produce large amounts of organic algal carbon: one pound of iron can fix eighty thousand pounds of carbon dioxide. Researchers also found that diatoms (which are rich in silicon) and coccolithophores (which are rich in calcium) are best at sequestering carbon; these heavier organisms sink more quickly and are more likely to take their carbon to the ocean floor without being eaten. By seeding ocean areas that have more dissolved silicon and calcium, geoengineers could more effectively increase carbon sequestration. Other factors are influential, too, such as the number of zooplankton in the waters and how quickly surface waters sink.

Given the numerous variables and the experimental difficulties, the resulting data vary widely. The Southern Ocean Iron Experiment (SOFeX) undertaken in 2002 between New Zealand and Antarctica found that although iron seeding did induce algal blooms, relatively little carbon was sequestered. On the other hand, Victor Smetacek, a marine biologist at the Alfred Wegener Institute for Polar and Marine Research, in conjunction with scientists from around the world, reported that at least half the algae encouraged by the 2004 European Iron Fertilization Experiment (EIFEX) would become sequestered — a rate thirty-four times the natural rate.

There's more than one way for iron seeding to skin the cat of climate change, however. Microalgae and macroalgae produce a sulfur compound known as dimethylsulfoniopropionate, or DMSP. This compound appears to help algae regulate their salinity or internal temperature and is a protective antioxidant. After algae excrete DMSP into seawater, bacteria break it down into gaseous dimethyl sulfide, or DMS. That ocean smell that drifts through your car's open window telling you you're almost at the beach? Part of what you detect is DMS. (Penguins follow the scent; for them, it signals algae, and algae mean fish.) Once lofted into the atmosphere, DMS further breaks down into microscopic particles small enough to stay suspended in the air. There they play an important role in the formation of clouds.

Water vapor that rises from the Earth's oceans doesn't spontaneously form clouds; first, it must condense on particles. DMS particles are an excellent size to act as cloud seeds (or, more formally, cloud condensation nuclei). While DMS from microalgae is not the only substance on which water vapor condenses—soot, dust, and ocean salt also act as nuclei—algae are major players in the cloud-making business. Satellite studies reveal that the greater the density of algae in the ocean, the more cloud cover there is. Because white clouds reflect light, they reduce the amount of solar energy that would otherwise be absorbed as heat by the oceans. During SOFeX, iron seeding increased DMS concentrations fourfold at the site. Some scientists posit that increasing DMS in even a small portion of the Southern Ocean would have a large cooling effect—but not a greenhouse gas–reducing effect—on our planet.

Still, iron seeding carries risks; there's a great deal we don't know about its possible collateral effects. The blooms we produce could include toxic algae. It's also possible that human-induced algal blooms could create dead zones, although none has done so yet. Because iron limits algae growth in the Southern Ocean, there is unused nitrogen and phosphorus floating in those waters. If we induce more algae, they would take up some portion of those nutrients. But what if, under normal circumstances, the nutrients eventually find their way to other oceans? Would iron seeding in the Southern Ocean increase productivity there, only to limit it elsewhere?

These are all concerns worth investigating. Nonetheless, we should recognize that for more than a hundred years humans have been drastically and unthinkingly altering our oceans. We've already annihilated edible species from cod to whales and reduced the average size of fish. We've already added massive amounts of nitrogen, phosphorus, and chemicals, as well as millions of tons of plastics to the water. We're already well on our way to completely destroying coral reefs and their inhabitants. And, because about 30 percent of the carbon we pour into the atmosphere becomes carbonic acid in the oceans, we've already increased the acidity of ocean water by an astounding 30 percent—the fastest change in ocean chemistry, according to the Smithsonian Institution, in the last fifty million years. The acidity thins the shells of mollusks and forces corals, coralline algae, and other carbonaceous organisms to spend more energy building their structures.

After an American company, Planktos, undertook an unauthorized and profit-oriented iron seeding experiment in 2012 (the company wanted to sell carbon credits for its iron seeding), the UN Environment Programme placed a moratorium on large-scale geoengineering experiments. But while intentional experiments have been put on hold, a natural experiment is underway. In 2017, scientists at Stanford reported on an annual summer algal bloom that turns two hundred thousand square miles of the Labrador Sea, off the southern coast of Greenland, a turquoise hue. They discovered that the bloom is fueled by iron particles eroded from rock and ferried into the sea on glacial meltwater. The natural bloom occurs with no negative impacts, or at least none that have been so far noted. The effects on carbon sequestration are as yet unknown, but that would make for interesting study.

Even if we do turn to iron seeding, no one suggests that algae could sequester more than 10 to 15 percent of the carbon dioxide humans emit annually, and it could be less. Still, if iron seeding works and doesn't cause concomitant harm, it could be one medicine in the varied pharmacopeia we're going to need to cure our climate.

Between 1950 and 2018, the CO_2 emissions generated by human activities rose from about 310 to more than 410 parts per million, the highest level for at least 800,000 years. If we continue to burn fossil fuels at the current rate in the next century, we will have doubled the current level of atmospheric carbon dioxide and engendered a temperature rise of 7 degrees Celsius (12.6 degrees Fahrenheit) or greater. According to a 2017 study published in *Nature Communications,* we'll be on our way to re-creating the climate that spawned the Azolla Event fifty million years ago.* What we could use is more research on iron seeding—under the control of internationally recognized scientific organizations—so we can make the best decisions, when the time comes.

In the long, long run, algae will help moderate the hothouse atmosphere we are creating, just as they have in the past. They will flourish in the new, shallow-water expanses created on our flooded continental shores, and they will feed on nitrogen and phosphorus flushed from the land by heavy rainfalls intrinsic to tropical climates. I can imagine that many thousands of years from now, some tribes of *Homo sapiens,* having survived drastic climate change, will be clustered along the northern shores of Canada and the Eurasian continent, shaded by banana and palm trees, warning their children not to swim in the soupy scum that lines the world's coastlines or play in the rotting seaweed washed up on the beaches. Maybe our distant descendants will recount stories of the ancient civilizations of the temperate and tropical latitudes, and wonder when algae will restore the environment in which they flourished.

*Even if we cease adding greenhouse gases to the atmosphere, complex feedbacks—like the increased methane release from melting permafrost that we are already observing—may well sustain the temperature rise.

Epilogue

It's easy to feel depressed, even defeated, by the continuing and seemingly unstoppable degradation of our environment. Carbon dioxide levels continue to climb; the oceans become ever more polluted, acidified, and depleted of corals and fish; arid regions expand in Africa while island nations vanish under the waves; species disappear at a rate that bodes a sixth global extinction. While there are few reasons to feel hopeful about the environment, the power of algae is one of them.

Algae certainly will plague us in ever-increasing numbers, but still they are a source of hope. We already know they can be harnessed to create fuel, plastics, animal feed, vitamins, protein, edible oils, and other useful products. On the environmental front, they can help remediate the waters we pollute. And, if push comes to calamitous shove, as the climate warms, iron-supplemented algae may help scrub an atmosphere overloaded with carbon dioxide.

The prospect of a carbon-neutral transportation fuel triggered my initial interest in algae. I now see that the designs of the first algae oil entrepreneurs were a bit like Da Vinci's drawings of flying machines: inspired, but impossible without more advanced technology. Today, to continue the analogy, I figure we are farther along than the Sopwith Camel but have yet to design a jet engine. New bioengineering techniques are immensely promising. We're just getting started on what algae can do.

Intriguing algae stories show up in my inbox every day. In May 2018, researchers led by Dr. Uwe Arnold at the Technical University of Munich published their work on a low-cost technique for converting algae grown with flue gas into lightweight, flexible-but-strong carbon fibers. The fibers, increasingly used in aircraft, vehicles, and construction to replace steel, aluminum, and concrete, are currently made from petroleum using processes that have a heavy carbon footprint. Moreover, carbon fibers are extremely durable, and can lock up CO_2 for millennia.

Other energy-related discoveries: The European Synchrotron Radiation Facility reported in *Science* in September 2017 that a team of international researchers have discovered an enzyme in *Chlorella* that allows it to convert some of its fatty acids into hydrocarbons using only light energy. And, in November 2018, Japanese researchers reported the discovery of a "switch" that controls an alga's ability to accumulate starches. By inactivating a certain enzyme, the researchers boosted the rate of accumulation tenfold. Both could lead to major advances in the biofuels arena.

I'm also keeping my eye on animal scientists who have discovered an entirely new reason to add a soupçon of seaweed to the feed of ruminant livestock, like cows, sheep, and goats. They do so not for the animals' nutrition, but to reduce the level of methane in the animals' burps and flatulence. The planet's 1.5 billion farm ruminants emit methane that, according to the FAO, is equivalent to 7.1 billion tons of carbon dioxide per year. That's nearly 15 percent of manmade carbon dioxide emissions—as much as we put into the atmosphere from burning transportation fuel. Silly as it sounds, cow belches and toots—or as scientists call them, ruminant enteric methane emissions—are a serious contributor to global warming.

The methane problem starts in an animal's first stomach chamber (its rumen) where bacteria ferment the tough carbohydrates in grasses and, in the process, produce the gas. For years, scientists have been looking for feed additives that would prevent the bacteria from producing the gas, but until recently, nothing worked, at least not for long. The cows or their bacteria adapted to every new additive and resumed puffing out the gas.

Then, about ten years ago, Robert Kinley and Rocky de Nys at James Cook University in Australia discovered that certain seaweeds added to cows' feed lowered their methane emissions, and they experimented with more than two dozen species to identify the most effective one. While all the seaweeds they tested had a positive effect, the high dosages required upset the bovines' digestion. Then the scientists tried *Asparagopsis taxiformis,* a macroalgae that looks something like a pink underwater fern and grows wild around Australia and other tropical and subtropical locations. In the lab, adding just a little of this seaweed – 2 percent of feed – to artificial cow stomachs reduces methane output so much that it becomes virtually undetectable.

The magic is performed by bromoform ($CHBr_3$), a compound in the seaweeds that protects them from bacterial infection. In cows' stomachs, the bromoform reacts with vitamin B_{12} to stop methane-producing bacteria from completing the last step in their creation of the gas. Kinley, de Nys, and their colleagues have since demonstrated that sheep fed a little *Asparagopsis* produced up to 85 percent less methane. Researchers are now analyzing results from experiments with cattle in Australia and dairy cows at the University of California, Davis. The preliminary reports from California are very encouraging: cows eating a diet that includes just 1 percent seaweed produce 50 percent less methane, and the reduction is immediate. What about the taste of the milk? In a blind test, 25 testers perceived no difference between samples from cows fed with and without seaweed. Good thing, because marketing milk with "a scent of the sea" would be a challenge.

Asparagopsis could be a big win for the environment. It also may be a win for animal farmers. Because cows expend energy to manufacture methane, animals fed *Asparagopsis* have more calories to direct to making proteins or milk, compounds that are more useful to their growth and our nutrition – and to a farmer's bottom line.

The seaweed could also be a win for developing countries. No one yet cultivates it, but the researchers expect it can be grown on long lines, much as carrageenan farmers in East Asia cultivate *Eucheuma cottonii.* A new seaweed industry would be a major economic benefit for Indonesia, the Philippines, and other tropical countries. And, if the seaweeds are grown where fertilizer runoff now feeds algal blooms and creates dead zones, they could help remediate those waters by soaking up the excess nutrients. It's early days yet, but *Asparagopsis* is promising.

Algae has long been a part of the human diet, and we'll be eating more of it in the future — and not just as seaweeds. I see more companies investing in algae as an animal protein substitute.

Growing animals for food is a remarkably inefficient way of nourishing people: Large animals eat many times the plant protein that they ultimately yield in meat. (According to David Pimentel, professor of ecology at Cornell University, if all the grain currently fed to livestock in the United States were instead consumed directly by humans, we could feed nearly 800 million people.) The good news is that the market for "faux meat" is expanding. Nestlé predicts that by 2020, $5 billion of plant-based meats will be sold in the US; Tyson Foods forecasts that by 2045, 20 percent of meat products sold in the US will be plant-based. While plants are currently the source of most of the protein in these products, algae — which produce protein even more efficiently and with far less impact on the environment — can be a part of the story, too.

Proteins are not just for eating. Every year, pharmaceutical companies create tens of billions of dollars' worth of "recombinant pharmaceuticals," protein-based drugs that are produced in laboratories by genetically altered cells. The cells — once primarily *E. coli* bacteria but increasingly mammalian cells — have added genes that instruct their natural protein-making machinery to express foreign proteins. These recombinant drugs include vaccines and medications that treat cancer, hormone deficiencies, and autoimmune and viral diseases. Some 400 recombinant protein-based drugs are currently used in medical treatments and nearly 1,500 are under development.

Neither bacterial nor mammalian cells, however, are ideal factories for making all proteins. While bacteria are easily engineered and efficiently produce many simple, small proteins, they lack the sophisticated assembly processes of eukaryotes. The problem is that proteins are made of amino acids that are bonded, coiled, and folded into complex three-dimensional structures, and bacteria can't perform all the steps to form the most elaborate shapes that humans need. On the other hand, while mammalian cells are capable of translating genetic instructions into the most sophisticated proteins, they are slow to reproduce and finicky about their living conditions, which makes large-scale production difficult and expensive. The average annual price of monoclonal antibody therapies made from mammalian cells is $100,000. And, using mammalian cells entails the risk of introducing cancer-causing gene sequences and infectious virus particles into the medication.

Several companies are looking to microalgae as an additional production platform. As eukaryotes, they have those advanced protein-making mechanisms that mammalian cells have, but they are less fussy, reproduce quickly, and they can't be infected with human pathogens. Bioengineers don't target the DNA in their nuclei, but in their chloroplasts; chloroplasts have the simple, circular chromosomes bequeathed to them by their cyanobacterial ancestors. In sum, in microalgae bioengineers get the best of all worlds: less complicated prokaryotic genes to alter together with sophisticated eukaryotic machinery to translate genetic instructions into complex proteins.

A few companies are now working to make cheaper, safer, algae-based recombinant drugs. Triton Algae Innovations, funded in part by a grant from the National Science Foundation, has created green algae that make a colostrum-like protein with the unique properties of human breast milk, which could be incorporated into infant formula. Lumen Bioscience, a Seattle startup supported with a grant from the National Institutes of Health, is working on an oral malaria vaccine that can tolerate higher temperatures, a useful property in countries where refrigeration is not always available. The list of recombinant protein medications in production grows daily; microalgae may be able to reduce their cost dramatically.

Algae. When you hear the word, you'll certainly still think of pond scum. Algae blooms are growing ever larger and lasting longer, threatening health and livelihoods. But I hope you'll now also think of the world's most powerful engines, the sun-powered green dynamos that constantly convert a noxious gas and water into the valuable stuff of life. I hope that from time to time you'll recall (using your algae-bolstered brain) how algae created — and constantly refresh — our oxygen-rich atmosphere. Remember that every fish in the sea depends on algae and that every plant on land is actually a sophisticated alga.

Eat heart-healthy salmon and slip a little seaweed into stews and soups to boost their flavor and nutrition. Be tickled that seaweed is in your ice cream and chocolate milk, and rest assured that carrageenan isn't dangerous. When you're shopping for fruits and vegetables, contemplate the seaweed biostimulants that brought them to market. Relish the prospect of cheaper medications produced by algae; look for less-polluting plastics and protein substitutes. Hope that entrepreneurs can raise enough *Asparagopsis* to reduce ruminants' methane emissions.

When will we read "Algae inside!" on gas pumps? Not soon enough, but it's primarily a question of price. Scientists and engineers will be making algae oil less expensive, but we need to level the economic playing field by incorporating the cost of fossil fuels' damage into their prices. Otherwise, algae fuel will be blocked from the marketplace — at great cost to all of us. Even if we replace only jet fuel, we'll have done the planet a great favor.

There will be no single, quick fix to our intertwined environmental crises, and it will take many small measures that have a significant effect in the aggregate. A 10 percent reduction in carbon dioxide emissions here; a 15 percent reduction in methane emissions there; a 5 percent drawdown of atmospheric carbon in the future — pretty soon you're talking about real environmental relief.

Algae: They created us, sustain us, and, if we're both clever and wise, they can help save us.

Acknowledgments

I could not have written this book without the help of dozens of scientists, entrepreneurs, and academics who generously shared their knowledge of all things algae. Let me highlight a few people who were especially generous with their time: Shelly Benson, Stephen Cunnane, Eun Kyoung Hwang, Jonathan Williams, Larch and Nina Hanson, Tollef Olson, Amha Belay, Jean-Paul Deveau, Jeff Hafting, Franklin Evans, Yonathan Zohar, Ryan Hunt, Craig Behnke, Christopher Yohn, Steve Mayfield, Paul Woods, Ed Legere, Jonathan Wolfson, Jill Kauffman Johnson, Nathalia Castro, Francesca Virdis, and Allen Stewart.

I am exceedingly grateful to Shanthi Chandrasekar, the gifted and versatile artist who provided the elegant drawings in the book.

I am no chef, so I give heartfelt thanks to Naomi Gibbs, Elisa Gobbo, and Cynthia Schollard for helping me out with the recipes in the appendix. I also am in debt to Mario Gobbo, Brittany Boser, Alan Kassinger, Ben Frank, and Terez Shea-Donohoe for checking my math and science. (Whatever errors remain are mine alone.) Thank you to Amy Panitz for accompanying me to Wales and cheerfully eating lots of seaweed with me. Special thanks to Steve Edelson for helping me learn to scuba dive, and to Marjorie Frank for going with me to a seaweed cooking class.

Another heartfelt thank-you to Naomi Gibbs for editing and shaping the manuscript in countless beneficial ways. Thanks to Lisa Sacks Warhol and Jennifer Freilach for their sharp pencils and perceptive questions. Gratitude, once again, to literary agent Michelle Tessler for her wise counsel.

As always, I am grateful to my husband, Ted, for indulging my unquenchable interest in my subject and my predilection for sharing (oversharing?) every fascinating fact and experience.

Appendix

Recipes

Here are some good recipes to get you started, but you can find many more on the websites of the seaweed companies mentioned in this book. There are seaweed cookbooks, too. Prannie Rhatigan's *Irish Seaweed Kitchen* is a good place to start. Dried seaweeds are usually available at grocery stores, and a wide variety can be found online.

Jill Burns's Tempura Sea Vegetable Medley

When I decided to cook with seaweed, it seemed that a class would be helpful. I found Cooking with Vegetables from the Sea at the Natural Gourmet Institute in Manhattan and convinced my friend Marjorie, who lives in Brooklyn, to go with me. I was hoping to find dishes with a significant component of seaweed, which were easy enough to prepare that I could add them to my limited repertoire. The following recipe is one that we cooked under the guidance of chef Jill Burns.

TEMPURA:

1½ cups whole wheat pastry flour
1 tablespoon garlic powder
¾ tablespoon arrowroot flour (or cornstarch)
½ cup seltzer water plus ½ cup water
1¾ cups loosely packed arame seaweed
2 tablespoons sesame oil
5 shallots, thinly sliced
1¼ cups water
2 small carrots, thinly sliced on the diagonal

1/4 cup soy sauce (or tamari)
4–5 cups canola oil

DIPPING SAUCE:

1/4 cup soy sauce (or tamari)
1/2 cup water
3 tablespoons lemon juice
2 tablespoons ginger juice (or 2 tablespoons minced ginger)

1. Whisk together the flour, garlic powder, and arrowroot in a bowl. Stirring well, add seltzer and water, creating a medium-thick batter. Cover and chill for 30–40 minutes. Batter will thicken as it chills. (You may need to add more water later if it is too thick.)
2. Prepare the dipping sauce by combining the soy sauce, water, lemon juice, and ginger juice. Pour into small bowls and set aside.
3. Rinse the arame in a bowl with cool water. Cover with water and soak for 5 minutes. Drain; rinse with cold water. Set aside.
4. Heat 2 tablespoons sesame oil in a skillet. Add the shallots. Sauté until golden brown, about 5 minutes. Add the arame, 1 1/4 cups water, and 2 tablespoons of the soy sauce. Bring to a boil, lower the heat, and simmer uncovered for 10 minutes. Then add the carrots and 2 more tablespoons of soy sauce. Cook until all liquid is evaporated. Transfer to a bowl to cool.
5. Heat the canola oil in a wok or large skillet over medium heat. Add the batter into the bowl of arame and vegetables. Stir gently.
6. Using a 1/4 cup or smaller measure, quickly scoop some of the mixture and drop into the hot oil. Fry until golden, about 3 minutes on each side. Fry 5 or 6 pieces at a time, avoiding overcrowding. Drain on paper towels and serve immediately with dipping sauce.

Fruit and Spirulina Smoothie

If you'd like to boost your intake of vitamin A and iron, add spirulina to a smoothie. It's always a trick to camouflage the taste of dried spirulina, but here's a recipe that works.

1 ripe banana
1½ cups of mixed frozen fruit
2 teaspoons spirulina powder
½ cup fresh baby spinach
½ to 1 cup water, soy, or almond milk
honey, to taste
a handful of blueberries for garnish

Add the first four ingredients plus ½ cup of liquid into the bowl of a blender. Blend. Add more liquid to achieve the consistency you prefer; add honey as desired. Garnish with blueberries for a dramatic contrast with the deep green smoothie.

Basic Miso Soup

Although wakame is almost always a part of miso soup, other vegetables—including cabbage, carrots, spring onions, mushrooms, and yams—can be added, and they boost the soup's vitamin content. Add clams and you will increase the protein content. The rice side dish is sometimes topped with a soy-pickled, fried, or poached egg or egg yolk. The Japanese often eat the rice by laying a strip of flavorful ajitsuke nori on top, then pinching the short ends with chopsticks to pick up a portion. Fortunately, Japanese short-grain rice is a little starchier and stickier than its long-grain cousins, which makes it easier to manipulate.

Miso soup for breakfast is easy, and, with rice and egg, more filling—and far more complex in terms of taste—than cereal. The one thing you don't want to do is boil the broth; a near simmer is as hot as you should go. I started out using white miso paste (it's actually a beige color), but found red miso more flavorful. Fresh tofu is a delight; while it doesn't have much taste, the pillowy texture is engaging, and a half-cup provides 20 percent of the recommended daily amount of protein. The only downside: I can't manage chopsticks while reading the newspaper!

4 cups water
one 4-inch piece of kombu (or *Saccharina latissima*)
¾ teaspoon bonito powder
1 teaspoon dried wakame (or ¼ cup flash-frozen)
6 ounces fresh tofu, cut into chunks
3 tablespoons white or red miso paste
¼ cup chopped scallions (or other ingredients as above)

1. Heat 4 cups of water in a large pot over low heat. Add kombu and cook until the water begins to simmer. Stir in bonito powder.
2. Add the wakame to a small bowl of water and set aside to reconstitute.
3. Remove pot from heat. Let the broth — called dashi — sit uncovered for 5 minutes.
4. Remove the kombu and reheat the dashi over medium heat, but do not boil. Drain the wakame and add it and the tofu to the pot.
5. Remove about a cup of the dashi to a small bowl and whisk in the miso paste. Add the mixture to the pot and stir to combine.
6. Add scallions. Serve with short-grain rice and ajitsuke nori strips.

Naomi's Spring Sea-Vegetable Tart in a Sesame Whole-Wheat Crust

This recipe is more complex, but the results are divine.

CRUST:

1 cup whole wheat flour
⅓ cup white flour
½ teaspoon salt (I used dulse salt for a
  bit of extra seaweed flavor)
½ teaspoon baking powder
1 tablespoon white sesame seeds
1 tablespoon black sesame seeds
⅓ cup sesame oil
⅓ cup water
1 tablespoon rice vinegar

FILLING:

1 pound extra-firm silken tofu
¼ teaspoon salt
2 cloves garlic
2 tablespoons soy sauce
1½ tablespoons miso paste
3 teaspoons rice vinegar
1 tablespoon freshly grated ginger
ground black pepper, to taste
1 teaspoon sesame oil
1 yellow onion, sliced into crescents
5–6 scallions, chopped
1 cup chopped spinach leaves
¾ cup chopped asparagus
2 tablespoons dried arame, soaked for 10 minutes
  in warm water, drained, and chopped
2 tablespoons fresh chives, chopped

To make the crust, mix flours, salt, baking powder, and both kinds of sesame seeds together in a bowl. In a separate bowl, mix together ⅓ cup oil, water, and vinegar. Pour into the dry ingredients and mix until integrated. Shape into a flat disc, cover, and refrigerate for at least an hour. When you're ready to bake, preheat the oven to 350°F. Roll out the dough with a rolling pin and put it into a tart pan. Poke holes in it with a fork and set in the oven. Bake for about 10 minutes, until it's starting to turn golden brown, then set aside to cool before putting the filling in. Do not turn off the oven.

Meanwhile, put the tofu, salt, garlic, soy sauce, miso paste, rice vinegar, ginger, and black pepper into the bowl of a food processor. Puree until you have a smooth, uniform mixture. Set aside.

Heat the remaining teaspoon of sesame oil in a pan, then add the sliced onions. Cook on medium to low heat until they're just turning brown, 10 to 15 minutes. Set them in the bowl with the tofu mixture, and add the scallions, spinach, asparagus, arame, and chives. Mix together until all the ingredients are evenly distributed, then pour into prepared tart shell. Set into the oven, still at 350°F, and bake for about an hour, or until the tart filling has set. Let cool for 10 to 20 minutes before serving.

Elisa's Dulse and Cheddar Scones

Dulse (*Palmaria*) is a beautiful dark rose- to burgundy-colored seaweed that grows abundantly on the northern coasts of the Pacific and Atlantic oceans. Its fronds are about 18 inches long and a just few inches wide.

In the year 600, the monks of St. Columba, in Scotland, noted that people ate dulse, and no doubt it had been on the menu for a long time. In an article titled "Purple Shore" in *Household Words*, a magazine edited by Charles Dickens, an anonymous author wrote in 1856 that the fishermen in the region pressed dulse "between two red-hot irons, which makes it taste like roasted oysters." Recalling childhood holidays in Aberdeen, the author remembers how, often, more than a dozen "dulse-wives" would be selling the seaweed:

> Of all the figures on the Castlegate, none were more picturesque than the dulse-wives. They sat in a row on little wooden stools, with their wicker creels placed before them on the granite paving stones. Dressed in clean white mutches, or caps, with silk-handkerchiefs spread over their breasts, and blue stuff wrappers and petticoats, the ruddy and sonsie [healthy] dulse-women looked the types of health and strength . . . Many a time, where my whole weekly income was a halfpenny, a Friday's bawbee [silver coin], I have expended it on dulse, in preference to apples, pears, blackberries, cranberries, strawberries, wild peas and sugar-sticks.

Dulse was generally eaten raw, the author reports, or used to season oat or wheat bread.

I ordered dried dulse from Maine Coast Sea Vegetables, both regular and applewood-smoked. I rehydrated it and nibbled it uncooked to see if I would have spent my halfpenny on dulse or blackberries. While I'd have bought the blackberries, I found the smoked dulse intriguing. It has a strong Scotch whisky flavor, and would be a treat on a tray of hors d'oeuvres along with cured Greek olives, sharp cheese, and other savory nibbles. And, dulse is a wonderful addition to cheese scones.

CHEF'S TIP:

Dulse flakes are dried and quite stiff. If they are added as an ingredient in baked goods, it is a good idea to first rinse the flakes quickly in a sieve under tepid water to give them a bit of moisture.

2 cups self-rising flour
¼ teaspoon salt
⅓ cup butter, chilled
2 tablespoons dried dulse, toasted briefly
and crumbled into flakes
1½ cups sharp cheddar cheese, grated
⅓ cup milk
2 large eggs, beaten

1. Preheat the oven to 425°F.
2. Sift the flour with the salt into a bowl.
3. With a fork or two knives, mix in the butter until the mixture resembles coarse crumbs. Add the dulse flakes and about ⅔ of the cheddar cheese and mix together.
4. In a cup or small bowl, stir together the milk and the eggs. Add to the flour mixture and incorporate briefly.
5. Turn dough onto a floured board and flatten it with your hands. Sprinkle with the rest of the cheddar. Cut into 16 squares and transfer them onto the baking sheet.
6. Bake in the middle of the oven for 10 to 17 minutes, depending on the size of the scones. Bake until golden and cooked through.

Welsh Cockles and Laver Stew

This is a variation of a traditional Welsh dish eaten at breakfast or dinner. Fresh cockles are difficult to find, so I substituted a canned product. Fresh steamed littleneck clams or mussels would work, too, but cockles are especially sweet. I used dried laver from Pembrokeshire Beach Food Company, but nori works just as well. The dish is rich; I serve it as an appetizer or for brunch, with a green salad and dark bread.

2½ tablespoons butter
½ onion, finely chopped
8 strips cooked bacon, chopped
4–5 tablespoons flour
14 ounces or more cream (or half-and-half)
2 tablespoons dried laver or nori, crumbled

pinch of pepper (white preferred)
8 ounces cockles, cooked or canned (or mussels or littleneck clams)
frozen puff pastry, thawed overnight in the refrigerator
1 egg, beaten

1. Preheat the oven to 350°F.
2. Melt the butter in a pan and cook the diced onion until slightly caramelized.
3. Add chopped, cooked bacon.
4. Add the flour and stir for a few minutes.
5. Slowly pour in cream until you have a fairly thick, smooth sauce.
6. Add the crumbled seaweed, a pinch of pepper, and cooked or canned cockles, mussels, or clams.
7. Place the mixture in four ovenproof ramekins (about 5-ounce size).
8. Cover the top of each ramekin with a disk of puff pastry; cut a small hole in the middle. Brush with beaten egg.
9. Bake for about 25 minutes, or until the pastry is golden.

Hummus with Seaweed

2 cups of cooked or canned cannellini beans, (reserve liquid)
3 cloves garlic, crushed
⅓ cup tahini
¼ cup extra-virgin olive oil (or Thrive algae oil), plus a little extra for drizzling
2½ tablespoons fresh lemon juice
2 teaspoons cumin
½ to 1 teaspoon ground coriander
½ teaspoon salt
1 to 2 packages (about ⅓ ounce total) seaweed snacks, ground in a food processor into small flakes, or 1 tablespoon alaria or nori powder

1. Put all the ingredients in a food processor and pulse until you have the texture you like. Add bean cooking or can liquid as necessary to thin mixture.
2. Put in a serving dish, drizzle oil on top, and serve with pita wedges or crackers.

Irish Moss Blancmange

This recipe is from Fannie Farmer's *The Boston Cooking-School Cook Book*, 1918.

⅓ cup Irish moss
4 cups milk
¼ teaspoon salt
1½ teaspoons vanilla
banana, sliced thin, for garnish

Soak moss 15 minutes in cold water to cover, drain, pick over, and add to milk; cook in double boiler 30 minutes. The milk will seem but little thicker than when put on to cook, but if cooked longer, blancmange will be too stiff. Add salt, strain, flavor with vanilla, re-strain, and fill individual molds previously dipped in cold water. Chill the molds, turn out on glass dish, surround with slices of banana, and place a slice on each mold. Serve with sugar and cream.

Chocolate Blancmange

This is Irish moss blancmange flavored with chocolate. Melt 1½ squares unsweetened chocolate, add ¼ cup sugar and ⅓ cup boiling water. Stir until perfectly smooth, adding to milk just before taking from fire. Serve with sugar and cream.

Selected and Annotated Bibliography

GENERAL

Barsanti, Laura, and Paolo Gualtieri. *Algae: Anatomy, Biochemistry, and Biotechnology*. CRC Press, 2014.

"FAO Fisheries & Aquaculture — Topics." Food and Agricultural Organization of the United Nations, 2018, www.fao.org/fishery/sofia/en. (The source for statistics on seaweed production around the globe.)

Graham, Linda E., et al. *Algae*. Benjamin Cummings, 2009.

"IPCC Special Report on Global Warming of 1.5 °C." UN Intergovernmental Panel on Climate Change, 2018, https://unfccc.int/topics/science/workstreams/cooperation-with-the-ipcc/ipcc-special-report-on-global-warming-of-15-degc.

Lembi, Carole A., and J. Robert Waaland. *Algae and Human Affairs*. Cambridge University Press, 1990.

McHugh, Dennis J. "A Guide to the Seaweed Industry." Food and Agricultural Organization of the United Nations, 2003, www.fao.org/docrep/006/y4765e/y4765e0b.htm.

"The National Climate Assessment." US Global Change Research Program, 2014, https://nca2014.globalchange.gov/.

Introduction

Nadis, Steve. "The Cells That Rule the Seas." *Scientific American*, vol. 289, no. 6, 10 Nov. 2003, pp. 52–53.

Pennisi, Elizabeth. "Meet the obscure microbe that influences climate, ocean ecosystems, and perhaps even evolution." *Science*, 9 Mar. 2017. (The number of algae in a drop of ocean water is based on 400,000 *Prochlorococcus* in one teaspoon.)

Walton, Marsha. "Algae: The Ultimate in Renewable Energy." www.cnn.com/2008/TECH/science/04/01/algae.oil/index.html. (For Kertz's oil production claims.)

Section I: In the Beginning

CHAPTER 1: POND LIFE

Artisans du Changement. "Takao Furuno: Des Canards dans la Rizière." www.youtube.com/watch?v=pqpEg45fp4I (English partial version at www.youtube.com/watch?v=SNR_3GeUoqI).

"The East Discovers Azolla." The Azolla Foundation, theazollafoundation.org/azolla/azollas-use-in-the-east.

Furuno, Takao. *The Power of Duck: Integrated Rice and Duck Farming*. Tagari Publications, 2001.

Wagner, Gregory M. "Azolla: A Review of Its Biology and Utilization." *The Botanical Review*, vol. 63, no. 1, 1997, pp. 1–26.

CHAPTER 2: SOMETHING NEW UNDER THE SUN

Biello, David. "The Origin of Oxygen in Earth's Atmosphere." *Scientific American*, 19 Aug. 2009.

Deamer, D. W. *First Life: Discovering the Connections between Stars, Cells, and How Life Began*. University of California Press, 2012.

Falkowski, Paul G. *Evolution of Primary Producers in the Sea*. Elsevier Academic Press, 2008.

——. *Life's Engines: How Microbes Made Earth Habitable*. Princeton University Press, 2017.

"Fast-Growth Cyanobacteria Have Allure for Biofuel, Chemical Production." *Lab Manager*, 28 July 2016, www.labmanager.com/news/2016/07/fast-growth-cyanobacteria-have-allure-for-biofuel-chemical-production. (Source for the speed of cyanobacterial reproduction.)

Feulner, Georg, et al. "Snowball Cooling after Algal Rise." *Nature News*, Nature Publishing Group, 27 Aug. 2015, www.nature.com/articles/ngeo2523.

Fortey, Richard. *Life: A Natural History of the First Four Billion Years of Life on Earth*. Alfred A. Knopf, 2000.

Hazen, Robert M. "Evolution of Minerals." *Scientific American*, Mar. 2010, www.scientificamerican.com/article/evolution-of-minerals.

——. *The Story of Earth: The First 4.5 Billion Years, from Stardust to Living Planet*. Penguin Books, 2013.

Lane, Nick. *Life Ascending: The Ten Great Inventions of Evolution*. W. W. Norton, 2010.

——. *Oxygen: The Molecule That Made the World*. Oxford University Press, 2016.

Nisbet, E. G., and N. H. Sleep. "The Habitat and Nature of Early Life." *Nature*, vol. 409, no. 6823, Mar. 2001, pp. 1083–1091. (Source for how cyanobacteria contributed to the formation of carbonate rock. The article is also a good summary of early life evolution.)

Pennisi, Elizabeth. "Meet the obscure microbe that influences climate, ocean ecosystems, and perhaps even evolution." *Science*, 9 Mar. 2017. (On the dominance of the tiny *Prochlorococcus* cyanobacteria.)

Schopf, J. William. *Cradle of Life: The Discovery of Earth's Earliest Fossils*. Princeton University Press, 2001.

"Timeline of Photosynthesis on Earth." *Scientific American*, www.scientificamerican.com/article/timeline-of-photosynthesis-on-earth/.

Whitton, Brian A. *Ecology of Cyanobacteria II: Their Diversity in Space and Time*. Springer, 2013.

CHAPTER 3: ALGAE GET COMPLICATED

Keeling, P. J. "The Endosymbiotic Origin, Diversification and Fate of Plastids." *Philosophical Transactions of the Royal Society B: Biological Sciences*, vol. 365, no. 1541, 2 Feb. 2010, pp. 729–748.

Le Page, Michael. "Why Complex Life Probably Evolved Only Once." *New Scientist*, 21 Oct. 2010.

Porter, Susannah. "The Rise of Predators." *Geology*, GeoScienceWorld, 1 June 2011, https://pubs.geoscienceworld.org/gsa/geology/article/39/6/607/130647/the-rise-of-predators.

Rai, Amar N., et al. *Cyanobacteria in Symbiosis*. Springer, 2011.

"Scientists Discover Clue to 2 Billion Year Delay of Life on Earth." *Phys.org*, 26 Mar. 2008, https://phys.org/news/2008-03-scientists-clue-billion-year-life.html.

Stanley, Steven M. "An Ecological Theory for the Sudden Origin of Multicellular Life in the Late Precambrian." *Proceedings of the National Academy of Sciences of the United States of America*, vol. 70, no. 5 (May 1973), pp. 1486–1489.

Yong, Ed. "The Unique Merger That Made You (and Ewe, and Yew)." *Nautilus*, 6 Feb. 2014. (On how a threesome likely made the first eukaryote.)

CHAPTER 4: LAND HO, GOING ONCE

"Ancient 'Great Leap Forward' for Life in the Open Ocean." *Astrobiology Magazine*, 9 Mar. 2014. (Nitrogen fertilization spurs blossoming of eukaryotic life.)

Becker, Burkhard. "Snow Ball Earth and the Split of Streptophyta and Chlorophyta." *Trends in Plant Science*, vol. 18, no. 4, Apr. 2013, pp. 180–183.

Boraas, Martin E., et al. "Phagotrophy by a Flagellate Selects for Colonial Prey: A Possible Origin of Multicellularity." *Evolutionary Ecology*, vol. 12, no. 2, Feb. 1998, pp. 153–164.

Brocks, Jochen J., Amber J. M. Jarrett, et al. "The Rise of Algae in Cryogenian Oceans and the Emergence of Animals." *Nature*, vol. 548, 31 Aug. 2017, pp. 578–581. (Melting glaciers spur eukaryotes.)

Collen, J., et al. "Genome Structure and Metabolic Features in the Red Seaweed Chondrus Crispus Shed Light on Evolution of the Archaeplastida." *Proceedings of the National Academy of Sciences*, vol. 110, no. 13, 2013, pp. 5247–5252. (On why red algae didn't evolve to become land plants.)

Delaux, Pierre-Marc, et al. "Algal Ancestor of Land Plants Was Preadapted for Symbiosis." *PNAS,* National Academy of Sciences, 27 Oct. 2015. (For evidence that plants inherited their ability to signal mycorrhizae from blue-green algae.)

Devitt, Terry. "Ancestors of Land Plants Were Wired to Make the Leap to Shore." University of Wisconsin–Madison *News,* 5 Oct. 2015.

Frazer, Jennifer. "Why Red Algae Never Packed Their Bags for Land." *Scientific American Blog Network,* 13 July 2015, blogs.scientificamerican.com/artful-amoeba/why-red-algae-never-packed-their-bags-for-land.

Lewis, L. A., and R. M. McCourt. "Green Algae and the Origin of Land Plants." *American Journal of Botany,* vol. 91, no. 10, Oct. 2004, pp. 1535–1556.

Niklas, Karl J., et al. "The Evolution of Hydrophobic Cell Wall Biopolymers: From Algae to Angiosperms." *Journal of Experimental Botany,* vol. 68, no. 19, 9 Nov. 2017, pp. 5261–5269. (For how early land plants survived desiccation with a genetic inheritance from algae.)

Salminen, Tiina, et al. "Deciphering the Evolution and Development of the Cuticle by Studying Lipid Transfer Proteins in Mosses and Liverworts." *Plants,* vol. 7, no. 1, 2018, p. 6.

CHAPTER 5: LAND HO, GOING TWICE

Brodo, Irwin M., et al. *Lichens of North America.* Yale University Press, 2001.

Chen, Jie, et al. "Weathering of Rocks Induced by Lichen Colonization—a Review." *Catena,* vol. 39, no. 2, Mar. 2000, pp. 121–146.

"Common Bryophyte and Lichen Species: Cladina: Reindeer Lichens." Boreal forest.org, www.borealforest.org/lichens.htm. (On the taste of reindeer lichen pudding.)

Frazer, Jennifer. "The World's Largest Mining Operation Is Run by Fungi." *Scientific American Blog Network,* 5 Nov. 2015, blogs.scientificamerican.com/artful-amoeba/the-world-s-largest-mining-operation-is-run-by-fungi. (Ten percent of Earth's surface is covered by lichens.)

Gadd, Geoffrey. "Fungi, Rocks, and Minerals." *Elements,* 17 June 2017, elementsmagazine.org/2017/06/01/fungi-rocks-and-minerals. (On fungal deconstruction of rocks.)

"Lichen and the Organic Evolution from Sea to Land." *The Liquid Earth Blog,* School of Ocean Sciences, Bangor University, 2012–13, theliquidearth.org/2012/10/lichen-and-the-organic-evolution-from-sea-to-land.

Yuan, Xunlai, et al. "An Early Ediacaran Assemblage of Macroscopic and Morphologically Differentiated Eukaryotes." *Nature,* vol. 470, no. 7334, 17 Feb. 2011, pp. 390–393.

——. "The Lantian Biota: A New Window onto the Origin and Early Evolution of Multicellular Organisms." *Chinese Science Bulletin,* vol. 58, no. 7, Mar. 2012, pp. 701–707.

——. "Lichen-Like Symbiosis 600 Million Years Ago." *Science,* American Association for the Advancement of Science, 13 May 2005.

Section II: Glorious Food

CHAPTER 1: BRAIN FOOD

Abraham, Guy E. "The History of Iodine in Medicine: Part 1." Optimox, www.optimox.com/iodine-study-14.

Bradbury, Joanne. "Docosahexaenoic Acid (DHA): An Ancient Nutrient for the Modern Human Brain." *Nutrients*, vol. 3, no. 5, 2011, pp. 529–554.

Burgi, H., et al. "Iodine deficiency diseases in Switzerland one hundred years after Theodor Kocher's survey." *European Journal of Endocrinology*, vol. 123, no. 6, Dec. 1990, pp. 577–590.

"Chimpanzees fishing for algae with tools in Bakoun, Guinea." www.youtube.com/watch?v=qEk_sNYAyCo.

Cunnane, Stephen C., and Kathlyn M. Stewart. *Human Brain Evolution: The Influence of Freshwater and Marine Food Resources*. Wiley-Blackwell, 2010. (Source also for EQ figures.)

——. *Survival of the Fattest: The Key to Human Brain Evolution*. World Scientific, 2006.

Eckhoff, Karen M., and Amund Maage. "Iodine Content in Fish and Other Food Products from East Africa Analyzed by ICP-MS." *Journal of Food Composition and Analysis*, vol. 10, no. 3, 7 July 1997, pp. 270–282.

Erlandson, Jon M., et al. "The Kelp Highway Hypothesis: Marine Ecology, the Coastal Migration Theory, and the Peopling of the Americas." *The Journal of Island and Coastal Archaeology*, vol. 2, no. 2, 30 June 2007, pp. 161–174.

Gibbons, Ann, et al. "The World's First Fish Supper." *Science*, 1 June 2010, www.sciencemag.org/news/201%6/worlds-first-fish-supper.

"Human Faces Are So Variable Because We Evolved to Look Unique." *Phys.org*, 16 Sept. 2014, https://phys.org/news/2014-09-human-variable-evolved-unique.html.

Kitajka, K., et al. "Effects of Dietary Omega-3 Polyunsaturated Fatty Acids on Brain Gene Expression." *Proceedings of the National Academy of Sciences*, vol. 101, no. 30, 19 July 2004, pp. 10931–10936.

MacArtain, Paul, et al. "Nutritional Value of Edible Seaweeds." *Nutrition Reviews*, vol. 65, no. 12, 2008, pp. 535–543.

Marean, Curtis W. "The Transition to Foraging for Dense and Predictable Resources and Its Impact on the Evolution of Modern Humans." *Philosophical Transactions of the Royal Society B: Biological Sciences*, vol. 371, no. 1698, 2016, p. 20, 150, 239.

——. "When the Sea Saved Humanity." *Scientific American*, 1 Nov. 2012, www.scientificamerican.com/article/when-the-sea-saved-humanity-2012-12-07.

Tarlach, Gemma. "Did the First Americans Arrive Via a Kelp Highway?" *Discover Magazine* Blogs, 2 Nov. 2017, blogs.discovermagazine.com/deadthings/2017/11/02/first-americans-kelp-highway.

Venturi, Sebastiano. "Evolutionary Significance of Iodine." *Current Chemical Biology*, vol. 5, no. 3, 2011, pp. 155–162.

Vynck, Jan C. De, et al. "Return Rates from Intertidal Foraging from Blombos

Cave to Pinnacle Point: Understanding Early Human Economies." *Journal of Human Evolution,* 28 Jan. 2016, pp. 101–115.

Wisniak, Jaime. "The History of Iodine From Discovery to Commodity." *Indian Journal of Chemical Technology,* vol. 8, Nov. 2001, http://nopr.niscair.res.in/bitstream/123456789/22953/1/IJCT%208%286%29%20518-526.pdf.

Zhao, Wei, et al. "Prevalence of Goiter and Thyroid Nodules before and after Implementation of the Universal Salt Iodization Program in Mainland China from 1985 to 2014: A Systematic Review and Meta-Analysis." *PLOS One,* 14 Oct. 2014, https://journals.plos.org/plosone/article?id=10.1371/journal.pone.0109549.

CHAPTER 2: SEAWEED SALVATION

Cian, Raúl, et al. "Proteins and Carbohydrates from Red Seaweeds: Evidence for Beneficial Effects on Gut Function and Microbiota." *Marine Drugs,* vol. 13, no. 8, 20 Aug. 2015, pp. 5358–5383.

Crawford, Elizabeth. "As Seaweed Snacks Gain Popularity, They Present a Chance to Get in at Ground Level, Expert Says." *foodnavigator-usa.com,* 29 Sept. 2015, www.foodnavigator-usa.com/Article/2015/09/30/Seaweed-gains-popularity-presenting-a-chance-to-get-in-at-ground-level.

Drew, Kathleen M. "Conchocelis-Phase in the Life-History of Porphyra umbilicalis (L.) Kütz." *Nature,* vol. 164, 29 Oct. 1949, p. 748–749. (On Drew-Baker's discovery of the reproductive cycle of *Porphyra.*)

Fitzgerald, Ciarán, et al. "Heart Health Peptides from Macroalgae and Their Potential Use in Functional Foods." *Journal of Agricultural and Food Chemistry,* vol. 59, no. 13, 2011, pp. 6829–6836.

Hehemann, Jan-Hendrik, et al. "Transfer of Carbohydrate-Active Enzymes from Marine Bacteria to Japanese Gut Microbiota." *Nature,* 8 Apr. 2010, www.nature.com/articles/nature08937.

Lund, J. W. G., et al. "Kathleen M. Drew D.Sc. (Mrs. H. Wright-Baker) 1901–1957." *British Phycological Bulletin,* vol. 1, no. 6, 1958, pp. iv–12.

Schaefer, Ernst, et al. "Plasma Phosphatidylcholine Docosahexaenoic Acid Content and Risk of Dementia and Alzheimer Disease." *Archives of Neurology,* vol. 63, no. 11, 2006, p. 1,545. (For evidence that high DHA content in the brain is linked to a reduction in Alzheimer's risk.)

CHAPTER 4: WELSHMEN'S DELIGHT

Morton, Chris. "Laverbread – The Story of a True Welsh Delicacy." *Bodnant Welsh Food,* 26 July 2013, www.bodnant-welshfood.co.uk/laverbread-2900/.html.

CHAPTER 5: A WAY OF LIFE

Yamamoto, S., et al. "Soy, isoflavones, and breast cancer risk in Japan." *Journal of the National Cancer Institute,* vol. 95, no. 12, 18 June 2003.

CHAPTER 6: FLASH!

Canfield, Clarke. "Boom in Urchin Harvest Flips Maine's Ecosystem." *Press Herald*, 26 Mar. 2013.

Kleiman, Dena. "Scorned at Home, Maine Sea Urchin Is a Star in Japan." *New York Times*, 3 Oct. 1990.

Sifton, Sam. "The Flavor Enhancer You Don't Need to Tell Anyone About." *New York Times*, 8 Feb. 2018. (On the benefits of cooking with dulse.)

CHAPTER 7: SPIRULINA

Renton, Alex. "If MSG Is So Bad for You, Why Doesn't Everyone in Asia Have a Headache?" *The Guardian*, Guardian News and Media, 10 July 2005. (Why MSG, the source of umami, does not cause headaches.)

Yang, Sarah. "New Study Finds Kelp Can Reduce Level of Hormone Related to Breast Cancer Risk." *UC Berkeley News*, Office of Public Affairs, 2 May 2005, www.berkeley.edu/news/media/releases/2005/02/02_kelp.shtml.

Section III: Practical Matters

CHAPTER 1: FEEDING PLANTS AND ANIMALS

BalterJul, Michael, et al. "Researchers Discover First Use of Fertilizer." *Science*, 15 July 2013, www.sciencemag.org/news/2013/07/researchers-discover-first-use-fertilizer. (On seaweeds as crop fertilizers.)

Barry, Kathleen A., et al. "Prebiotics in Companion and Livestock Animal Nutrition." In *Prebiotics and Probiotics Science and Technology*, edited by Dimitris Charalampopoulos and Robert A. Rastall, Springer, 2009.

Battacharyya, Dhriti, et al. "Seaweed Extracts as Biostimulants in Horticulture." *Scientia Horticulturae*, vol. 196, 2015, pp. 39–48.

Calvo, Pamela, et al. "Agricultural Uses of Plant Biostimulants." *Plant and Soil*, vol. 383, nos. 1–2, 2014, pp. 3–41.

Evans, F. D., and A. T. Critchley. "Seaweeds for Animal Production Use." *Journal of Applied Phycology*, vol. 26, no. 2, 10 Sept. 2013, pp. 891–899. (For data on the value of adding *Ascophyllum* to livestock diets.)

Flint, Harry J., et al. "The Role of the Gut Microbiota in Nutrition and Health." *Nature Reviews Gastroenterology & Hepatology*, vol. 9, no. 10, 4 Sept. 2012, pp. 577–589.

"The Importance of Seaweed across the Ages." *BioMara*, www.biomara.org/understanding-seaweed/the-importance-of-seaweed-across-the-ages.html.

Khan, Wajahatullah, et al. "Seaweed Extracts as Biostimulants of Plant Growth and Development." *Journal of Plant Growth Regulation*, vol. 28, no. 4, Dec. 2009, pp. 386–399.

Lloyd-Price, Jason, et al. "The Healthy Human Microbiome." *Genome Medicine*, vol. 8, no. 1, 27 Apr. 2016.

O'Sullivan, Laurie, et al. "Prebiotics from Marine Macroalgae for Human and

Animal Health Applications." Multidisciplinary Digital Publishing Institute, 1 July 2010, www.mdpi.com/1660-3397/8/7/2038.

Saad, N., et al. "An Overview of the Last Advances in Probiotic and Prebiotic Field." *LWT—Food Science and Technology*, vol. 50, no. 1, 21 May 2012, pp. 1–16.

"Tackling Drug-Resistant Infections Globally." *Review on Antimicrobial Resistance,* May 2016, https://amr-review.org/sites/default/files/160525_Final%20paper_with%20cover.pdf.

CHAPTER 2: IN THE THICK OF IT

Bixler, Harris J., and Hans Porse. "A Decade of Change in the Seaweed Hydrocolloids Industry." *Journal of Applied Phycology*, vol. 23, no. 3, 22 Apr. 2010, pp. 321–335.

Callaway, Ewen. "Lab Staple Agar Hit by Seaweed Shortage." *Algae World News,* 13 Dec. 2015, news.algaeworld.org/2015/12/lab-staple-agar-hit-by-seaweed-shortage.

Connett, David. "Why a Global Seaweed Shortage Is Bad News for Scientists." *The Independent,* 3 Jan. 2016.

Downs, C. A., et al. "Toxicopathological Effects of the Sunscreen UV Filter, Oxybenzone (Benzophenone-3), on Coral Planulae and Cultured Primary Cells and Its Environmental Contamination in Hawaii and the U.S. Virgin Islands." *SpringerLink,* 20 Oct. 2015, https://link.springer.com/article/10.1007%2Fs00244-015-0227-7.

"FAO/WHO Joint Expert Committee on Food Additives (JECFA) Releases Technical Report on Carrageenan Safety in Infant Formula." *PR Newswire,* June 2015.

Fernandes, Susana C. M., et al. "Exploiting Mycosporines as Natural Molecular Sunscreens for the Fabrication of UV-Absorbing Green Materials." *ACS Applied Materials & Interfaces,* vol. 7, no. 30, 13 July 2015, pp. 16558–165564.

Fleming, Derek, and Kendra Rumbaugh. "Approaches to Dispersing Medical Biofilms." *Microorganisms,* vol. 5, no. 2, 1 Apr. 2017, pp. 1–16. (Algal drugs that disrupt these living membranes are a hallmark of certain diseases, including cystic fibrosis.)

"Is Carrageenan Safe?" *Follow Your Heart,* followyourheart.com/is-carrageenan-safe.

Iselin, Josie. "The Hidden Life of Seaweed: How a Rockland Seaweed Factory Helped Create the Processed Foods We Know and Love Today." *Maine Boats Homes & Harbors,* 11 July 2016, maineboats.com/print/issue-129/hidden-life-seaweed.

Keane, Kaitlin. "Video: Irish Sea Mossers Recall 40 Years of Camaraderie at Reunion." *Patriot Ledger* (Quincy, MA), 25 Aug. 2008, www.patriotledger.com/article/20080825/News/308259948.

Lawrence, Karl, and Antony Young. "Your Sunscreen May Be Polluting the Ocean—But Algae Could Offer a Natural Alternative." *The Conversation,* 1 Sept. 2017, http://theconversation.com/your-sunscreen-may-be-polluting-the-ocean-but-algae-could-offer-a-natural-alternative-83261.

Martin, Roy E., and Brian Rudolph. "Sea Products: Red Algae of Economic Significance." *Marine & Freshwater Products Handbook*. Technomic, 2000.

McHugh, Dennis J. "A Guide to the Seaweed Industry." Food and Agricultural Organization of the United Nations, 2003, www.fao.org/docrep/006/y4765e/y4765e0b.htm.

McKim, James M., et al. "Effects of Carrageenan on Cell Permeability, Cytotoxicity, and Cytokine Gene Expression in Human Intestinal and Hepatic Cell Lines." *Food and Chemical Toxicology*, vol. 96, Oct. 2016, pp. 1–10.

Murphy, Barbara. *Irish Mossers and Scituate Harbour Village*. B. Murphy, 1980.

Pritchard, Manon F., et al. "A Low-Molecular-Weight Alginate Oligosaccharide Disrupts Pseudomonal Microcolony Formation and Enhances Antibiotic Effectiveness." *Antimicrobial Agents and Chemotherapy*, vol. 61, no. 9, 19 Aug. 2017. (On biofilm disruption.)

Romo, Vanessa. "Hawaii Approves Bill Banning Sunscreen Believed to Kill Coral Reefs." NPR, 2 May 2018, www.npr.org/sections/thetwo-way/2018/05/02/607765760/hawaii-approves-bill-banning-sunscreen-believed-to-kill-coral-reefs.

"The Seaweed Site: Information on Marine Algae." www.seaweed.ie/uses_general/carrageenans.php. (Good summary of the carrageenan controversy.)

Stoloff, Leonard. "Irish Moss—from an Art to an Industry." *Economic Botany*, vol. 3, no. 4, 1949, pp. 428–435.

"Take the Luck Out of Clear Beer with Irish Moss." American Homebrewers Association, www.homebrewersassociation.org/how-to-brew/take-the-luck-out-of-clear-beer-with-irish-moss.

Valderamma, Diego. *Social and Economic Dimensions of Carrageenan Seaweed Farming*. Food and Agricultural Organization of the United Nations, 2013, www.fao.org/3/a-i3344e.pdf.

Wagner, Lisa. "Chemicals in Sunscreen Are Harming Coral Reefs, Says New Study." National Public Radio, *The Two-Way*, 20 Oct. 2015, www.npr.org/sections/thetwo-way/2015/10/20/450276158/chemicals-in-sunscreen-are-harming-coral-reefs-says-new-study.

Watson, Duika Burges. "Public Health and Carrageenan Regulation: A Review and Analysis." *Journal of Applied Phycology*, vol. 20, no. 5, 2007, pp. 505–513.

Weiner, Myra L. "Food Additive Carrageenan: Part II: A Critical Review of Carrageenan in vivo Safety Studies." *Critical Reviews in Toxicology*, vol. 44, no. 3, 2014, pp. 244–269.

Yang, Guang, et al. "Photosynthetic Production of Sunscreen Shinorine Using an Engineered Cyanobacterium." *ACS Synthetic Biology*, vol. 7, no. 2, 2018, pp. 664–671.

CHAPTER 3: LAND HO, GOING THRICE

Abolofia, J., F. Asche, and J. E. Wilen. "The Cost of Lice: Quantifying the Impacts of Parasitic Sea Lice on Farmed Salmon." 21 April 2017, *Marine Resource Economics*, 32(3), 329–349.

Andrews, Ethan. "Norwegian Company to Build Large, Land-Based Salmon Farm in Belfast." *Press Herald,* 31 Jan. 2018.

Cruz-Suarez, Lucia Elizabeth, et al. "Shrimp/*Ulva* co-culture: A sustainable alternative to diminish the need for artificial feed and improve shrimp quality." *Aquaculture,* vol. 301, nos. 1–4, 23 Mar. 2010, pp. 64–68.

Elizondo-González, Regina, et al. "Use of Seaweed *Ulva Lactuca* for Water Bioremediation and as Feed Additive for White Shrimp *Litopenaeus Vannamei.*" *PeerJ,* 5 Mar. 2018, https://peerj.com/articles/4459.

Gui, Jian-Fang. *Aquaculture in China: Success Stories and Modern Trends.* John Wiley & Sons, 31 Mar. 2018. (See chapter 3.12, "Rabbitfish: An Emerging Herbivorous Marine Aquaculture Species.")

Managing Forage Fish — Recommendations from the Lenfest Task Force. Pew Charitable Trusts, 13 Apr. 2017, www.lenfestocean.org/en/news-and-publications/published-paper/managing-forage-fish-recommendations-from-the-lenfest-task-force.

"New Fish Farms Move from Ocean to Warehouse." Worldwatch Institute, 10 Apr. 2018, www.worldwatch.org/node/5718.

Parshley, Lois. "The Most Sustainable Way to Raise Seafood Might Be on Land." *Popular Science,* 22 Sept. 2015. (For Indiana's shrimp farms.)

CHAPTER 4: SEAWEED STUFF

Dungworth, David. "Innovations in the 17th-Century Glass Industry: The Introduction of Kelp (Seaweed) Ash in Britain." Association Verre Et Histoire, 14 June 2011.

Geyer, R., J. Jambeck, and K. Law. "Production, use, and fate of all plastics ever made." *Science Advances,* 19 July 2017, vol. 3, no. 7.

"The Importance of Seaweed across the Ages." *BioMara,* www.biomara.org/understanding-seaweed/the-importance-of-seaweed-across-the-ages.html.

"How Once-Popular Pool Halls Ushered in the Age of Plastic." 99% Invisible. *Slate,* 13 May 2015, www.slate.com/blogs/the_eye/2015/05/13/the_death_of_billiards_and_the_rise_of_plastic_on_99_invisible_with_roman.html.

Johnson, Samuel. "A Journey to the Western Isles of Scotland." Gutenberg.org, 5 Apr. 2005, digital.library.upenn.edu/webbin/gutbook/lookup?num=2064.

"Kelp Burning in Orkney." *Orkneyjar — The Heritage of the Orkney Islands,* orkneyjar.com/tradition/kelpburning.htm.

Rymer, Leslie. "The Scottish Kelp Industry." *Scottish Geographical Magazine,* vol. 90, Dec. 1974.

CHAPTER 5: ALGAE OIL

Aarhus University. "Hydrothermal Liquefaction — Most Promising Path to Sustainable Bio-Oil Production." *Phys.org,* 6 Feb. 2013, phys.org/news/2013-02-hydrothermal-liquefactionmost-path-sustainable-bio-oil.html.

Algae to Crude Oil: Million-Year Natural Process Takes Minutes in the Lab. Pacific Northwest National Laboratory, 17 Dec. 2013, www.pnnl.gov/news/release.aspx?id=1029. (For more on hydrothermal liquefaction.)

Barreiro, Diego López, et al. "Hydrothermal Liquefaction (HTL) of Microalgae for Biofuel Production: State of the Art Review and Future Prospects." *Biomass and Bioenergy*, vol. 53, 8 Feb. 2013, pp. 113–127.

"Crude Oil Prices — 70 Year Historical Chart." Macrotrends, 2018, www.macrotrends.net/1369/crude-oil-price-history-chart.

Davis, Ryan, et al. "Techno-Economic Analysis of Autotrophic Microalgae for Fuel Production." *Applied Energy*, vol. 88, no. 10, 17 May 2011, pp. 3524–3531, https://www.sciencedirect.com/science/article/pii/S0306261911002406. (The authors at the National Renewable Energy Laboratory calculated in 2011 that algae diesel would be $9.84 per gallon, including a 10 percent return.)

Dong, Tao, et al. "Combined Algal Processing: A Novel Integrated Biorefinery Process to Produce Algal Biofuels and Bioproducts." *Algal Research*, 18 Jan. 2016. (For a new process that captures sugars, lipids, and proteins to maximize biofuel production and lower its cost.)

Elliott, Douglas C., et al. "Hydrothermal Liquefaction of Biomass: Developments from Batch to Continuous Process." *Bioresource Technology*, vol. 178, 13 Oct. 2014, pp. 147–156. (On fast hydrothermal liquefaction.)

Gluck, Robert. *Q & A with Sapphire Energy's Mike Mendez — Part I*. https://biofuelsdigest.blogspot.com/2010/11/q-with-sapphire-energys-mike-mendez.html, 19 Nov. 2010.

——. *Q & A with Sapphire Energy's Mike Mendez — Part II*. Nov. 2010. (Formerly available online at biofuelsdigest.blogspot.com.)

Kumar, Ramanathan Ranjith, et al. "Lipid Extraction Methods from Microalgae: A Comprehensive Review." *Frontiers in Energy Research*, vol. 2, 8 Jan. 2015.

Li, Yan, et al. "A Comparative Study: The Impact of Different Lipid Extraction Methods on Current Microalgal Lipid Research." *Microbial Cell Factories*, BioMed Central, 24 Jan. 2014.

Mitra, Aditee. "The Perfect Beast." *Scientific American*, vol. 318, no. 4, 20 Apr. 2018, pp. 26–33, doi:10.1038/scientificamerican0418-26. (On the newly discovered ubiquity of mixotrophs.)

"Sapphire Press Release Extracts." *Sapphire Energy*. Sapphire.com.

Woody, Todd. "The U.S. Military's Great Green Gamble Spurs Biofuel Startups." *Forbes*, 25 Sept. 2012.

Yap, Benjamin H. J., et al. "Nitrogen Deprivation of Microalgae: Effect on Cell Size, Cell Wall Thickness, Cell Strength, and Resistance to Mechanical Disruption." *Journal of Industrial Microbiology & Biotechnology*, vol. 43, no. 12, 6 Oct. 2016, pp. 1671–1680. (For data on algae cell wall thickness.)

CHAPTER 6: THE ALGAE'S NOT FOR BURNING

Abdelhamid, A. S., et al. "Omega 3 fatty acids for the primary and secondary prevention of cardiovascular disease." ResearchGate, July 2018, www.re

searchgate.net/publication/326482538_Omega-3_fatty_acids_for_the_primary_and_secondary_prevention_of_cardiovascular_disease.

"Bon Appétit Management Company Adopts TerraVia's Innovative Algae Oils." BusinessWire, 31 Jan. 2017, www.businesswire.com/news/home/20170131005375/en/Bon-App%C3%A9tit-Management-Company-Adopts-TerraVias-Innovative.

Byelashov, Oleksandr A., and Mark E. Griffin. "Fish In, Fish Out: Perception of Sustainability and Contribution to Public Health." *Fisheries*, vol. 39, no. 11, 24 Nov. 2014, pp. 531–535. (For commentary on the decline of omega-3 oils in aquacultured fish.)

Cardwell, Diane. "For Solazyme, a Side Trip on the Way to Clean Fuel." *New York Times*, 22 June 2013.

Essington, T. E., et al. "Fishing amplifies forage fish population collapses." *PNAS*, 26 May 2015, https://doi.org/10.1073/pnas.1422020112.

Hage, Øystein, and Fiskeribladet Fiskaren. "Skretting Exec: Consumers Must Accept Lower Omega 3 Levels." *IntraFish*, 6 May 2016, www.intrafish.com/news/489562/skretting-exec-consumers-must-accept-lower-omega-3-levels.

"The Science—and Environmental Hazards—Behind Fish Oil Supplements." *Fresh Air*, Terry Gross interview with Paul Greenberg, author of *The Omega Principle: Seafood and the Quest for a Long Life and a Healthier Planet*, 9 July 2018, www.npr.org/2018/07/09/627229213/the-science-and-environmental-hazards-behind-fish-oil-supplements.

"The Use of Algae in Fish Feeds as Alternatives to Fishmeal." *The Fish Site*, 13 Nov. 2013, thefishsite.com/articles/the-use-of-algae-in-fish-feeds-as-alternatives-to-fishmeal.

White, Cliff. "Algae-Based Aqua Feed Firms Breaking down Barriers for Fish-Free Feeds." *Seafoodsource.com*, 6 Apr. 2017, www.seafoodsource.com/news/aquaculture/algae-based-aquafeed-firms-breaking-down-barriers-for-fish-free-feeds.

CHAPTER 7: ETHANOL

Abbasi, Jennifer. "Kill Switches for GMOs." *Scientific American*, vol. 313, no. 6, 17 Nov. 2015, p. 36. (On preventing the escape of modified organisms into the environment.)

Boettner, Benjamin. "Kill Switches for Engineered Microbes Gone Rogue." *Wyss Institute*, 21 May 2018, wyss.harvard.edu/kill-switches-for-engineered-microbes-gone-rogue.

Mumm, Rita H, et al. "Land Usage Attributed to Corn Ethanol Production in the United States." *Biotechnology and Biofuels*, 12 Apr. 2014.

CHAPTER 8: THE FUTURE OF ALGAE FUEL

Bittman, Mark. "Is Natural Gas 'Clean'?" *New York Times Opinionator*, 24 Sept. 2013, opinionator.blogs.nytimes.com/2013/09/24/is-natural-gas-clean.

"Crude Oil Prices—70 Year Historical Chart." Macrotrends, 2018, www.macrotrends.net/1369/crude-oil-price-history-chart.

"Facts and Figures." Air Transport Action Group, www.atag.org/facts-figures.html. (For aviation's contribution to total carbon dioxide emissions.)

Gilson, Dave, and Benjy Hansen-Bundy. "How Big Oil Clings to Billions in Government Giveaways." *Mother Jones*, 24 June 2017.

Guglielmi, Giorgia. "Methane Leaks from US Gas Fields Dwarf Government Estimates." *Nature*, vol. 558, no. 7711, 28 June 2018, pp. 496–497.

Hanson, Chris. "Algae Tricked into Staying up Late to Produce Biomaterials." *Biomassmagazine.com*, 19 Nov. 2013, biomassmagazine.com/articles/9708/algae-tricked-into-staying-up-late-to-produce-biomaterials.

Kim, Hyun Soo, et al. "High-Throughput Droplet Microfluidics Screening Platform for Selecting Fast-Growing and High Lipid-Producing Microalgae from a Mutant Library." *Freshwater Biology*, 27 Sept. 2017.

Lane, Jim. "A Breakthrough in Algae Harvesting." *Biofuels Digest*, 21 Aug. 2016, www.biofuelsdigest.com/bdigest/2016/08/21/a-breakthrough-in-algae-harvesting.

"Proof That a Price on Carbon Works." *New York Times*, 19 Jan. 2016.

Salisbury, David. "Tricking Algae's Biological Clock Boosts Production of Drugs, Biofuels." Vanderbilt University, 7 Nov. 2013, news.vanderbilt.edu/2013/11/07/algaes-clock-drugs-biofuels.

Stockton, Nick. "Fattened, Genetically Engineered Algae Might Fuel the Future." *Wired*, Conde Nast, 20 June 2017.

Waller, Peter, et al. "The Algae Raceway Integrated Design for Optimal Temperature Management." *Biomass and Bioenergy*, vol. 46, 11 Aug. 2012, pp. 702–709.

"World Jet Fuel Consumption by Year." *IndexMundi*, 2013, www.indexmundi.com/energy/?product=jet-fuel.

Section IV: Algae and the Changing Climate

CHAPTER 1: GADZOOX

Dubinsky, Zvy, and Noga Stambler. *Coral Reefs: An Ecosystem in Transition*. Springer, 2014.

Harvey, Martin. "Coral Reefs: Importance." World Wildlife Fund, wwf.panda.org/our_work/oceans/coasts/coral_reefs/coral_importance. (Data on the economic importance of coral reefs.)

Klein, Joanna. "In the Deep, Dark Sea, Corals Create Their Own Sunshine." *New York Times*, 8 July 2017. (On how corals create light for their zoox.)

Thurber, Rebecca Vega, et al. "Macroalgae Decrease Growth and Alter Microbial Community Structure of the Reef-Building Coral, *Porites Astreoides*." *PLOS One*, 5 Sept. 2012.

CHAPTER 2: SAVING THE REEFS?

Apprill, Amy M., and Ruth D. Gates. "Recognizing Diversity in Coral Symbiotic Dinoflagellate Communities." *Molecular Ecology*, vol. 16, no. 6, 2006, pp. 1127–1134.

Berkelmans, R., and M. J. H. van Oppen. "The Role of Zooxanthellae in the Thermal Tolerance of Corals: A 'Nugget of Hope' for Coral Reefs in an Era of Climate Change." *Proceedings of the Royal Society B: Biological Sciences*, vol. 273, no. 1599, 2006, pp. 2305–2312.

Cai, Wenju, et al. "Increasing frequency of extreme El Niño events due to greenhouse warming." *Nature Climate Change*, 19 Jan. 2014, vol. 4, pp. 111–116.

Kline, David I., and Steven V. Vollmer. "White Band Disease (Type I) of Endangered Caribbean Acroporid Corals Is Caused by Pathogenic Bacteria." Nature.com, *Scientific Reports*, vol. 1, no. 1, 14 June 2011, www.nature.com/articles/srep00007.

Leibach, Julie. "Coral Sperm Banks: A Safety Net for Reefs?" *Science Friday*, 1 June 2016, www.sciencefriday.com/articles/coral-sperm-banks-a-safey-net-for-reefs/.

Little, A. F., et al. "Flexibility in Algal Endosymbioses Shapes Growth in Reef Corals." *Science*, vol. 304, no. 5676, 4 June 2004, pp. 1492–1494.

Morris, Emily, and Ruth D. Gates. "Functional Diversity in Coral-Dinoflagellate Symbiosis." *PNAS*, National Academy of Sciences, 8 July 2008.

Oppen, Madeleine J. H. van, et al. "Building Coral Reef Resilience through Assisted Evolution." *PNAS*, National Academy of Sciences, 24 Feb. 2015.

Pala, Chris. "Bonaire: The Last Healthy Coral Reef in the Caribbean." *The Ecologist*, 17 Nov. 2017, theecologist.org/2011/jan/04/bonaire-last-healthy-coral-reef-caribbean.

Putnam, H. M., and R. D. Gates. "Preconditioning in the Reef-Building Coral *Pocillopora damicornis* and the Potential for Trans-Generational Acclimatization in Coral Larvae under Future Climate Change Conditions." *Journal of Experimental Biology*, vol. 218, no. 15, 2015, pp. 2365–2372. (On building tolerance of corals in labs and research on passing down epigenetic changes.)

Sampayo, E. M., et al. "Bleaching Susceptibility and Mortality of Corals Are Determined by Fine-Scale Differences in Symbiont Type." *Proceedings of the National Academy of Sciences*, vol. 105, no. 30, 2008, pp. 10444–10449. (Research on the importance of *Symbiodinium* clades in coping with climate change.)

Thurber, Rebecca Vega, et al. "Macroalgae Decrease Growth and Alter Microbial Community Structure of the Reef-Building Coral, *Porites astreoides*." *PLOS One*, 5 Sept. 2012, https://journals.plos.org/plosone/article?id=10.1371/journal.pone.0044246.

CHAPTER 3: A PLAGUE UPON US

"The Algae Is Coming, But Its Impact Is Felt Far from Water." *NPR*, 11 Aug. 2013, www.npr.org/2013/08/11/211130501/the-algae-is-coming-but-its-impact-is-felt-far-from-water.

Bargu, Sibel, et al. "Mystery behind Hitchcock's birds." *Nature Geoscience*, vol. 5, no. 1, 2011, pp. 2–3.

Dodds, Walter. "Eutrophication of US Freshwaters: Analysis of Potential Economic Damages." *Environmental Science & Technology*, vol. 43, no. 1, 12 Nov. 2008, pp. 12–19. (Source for $2.2 billion damage estimate.)

Elsken, Katrina. "There's More to the Story: Invasion of the Algae Megabloom." *Okeechobee News,* 22 July 2016, okeechobeenews.net/lake-okeechobee/theres-story-invasion-algae-megabloom.

Gulf Shrimp Prices Reveal Hidden Economic Impact of Dead Zones. Duke University, Nicholas School of the Environment, 30 Jan. 2017, nicholas.duke.edu/about/news/gulf-shrimp-prices-reveal-hidden-economic-impact-dead-zones.

Hauser, Christine. "Algae Bloom in Lake Superior Raises Worries on Climate Change and Tourism." *New York Times,* 29 Aug. 2018, https://www.nytimes.com/2018/08/29/science/lake-superior-algae-toxic.html.

Lyn, Cheryl. "Dead Zones Spreading in World Oceans." *OUP Academic,* Oxford University Press, *BioScience,* vol. 55, no. 7, 1 July 2005, pp. 552–557.

Milstein, Michael. "NOAA Fisheries mobilizes to gauge unprecedented West Coast toxic algal bloom." Northwest Fisheries Science Center, June 2015, www.nwfsc.noaa.gov/news/features/west_coast_algal_bloom/index.cfm.

Nobel, Mariah. "Utah Lake Reopens as Algal Threat Subsides." *Salt Lake Tribune,* 30 July 2016, updated 4 Jan. 2017.

"Toxic Algal Blooms behind Klamath River Dams Create Health Risks Far Downstream." *Life at OSU,* 16 June 2015, today.oregonstate.edu/archives/2015/jun/toxic-algal-blooms-behind-klamath-river-dams-create-health-risks-far-downstream.

Zimmer, Carl. "Cyanobacteria Are Far from Just Toledo's Problem." *New York Times,* 20 Dec. 2017.

(See also sources in Chapter 5.)

CHAPTER 4: CLEANUP

Adey, Walter, and Karen Loveland. *Dynamic Aquaria: Building and Restoring Living Ecosystems.* Academic Press, 2007.

"Algae: A Mean, Green Cleaning Machine." USDA *AgResearch Mag,* May 2010, agresearchmag.ars.usda.gov/2010/may/algea.

Calahan, Dean, and Ed Osenbaugh. "Algal Turf Scrubbing: Creating Helpful, Not Harmful, Algal 'Blooms.'" Science Trends, 25 May 2018, sciencetrends.com/algal-turf-scrubbing-creating-helpful-not-harmful-algal-blooms.

Staletovich, Jenny. "Lake Okeechobee: A Time Warp for Polluted Water." *Orlando Sentinel,* 20 Aug. 2016, www.orlandosentinel.com/news/environment/os-ap-okeechobee-polluted-water-20160820-story.html.

———. "Massive and Toxic Algae Bloom Threatens Florida Coasts with Another Lost Summer." *Miami Herald,* 29 June 2018, updated 7 Aug. 2018.

Warrick, Joby. "Large 'Dead Zone' Signals Continued Problems for the Chesapeake Bay." *Washington Post,* 31 Aug. 2014.

CHAPTER 5: MAKING MONSTERS

"The Algae Is Coming, But Its Impact Is Felt Far from Water." NPR, 11 Aug. 2013, www.npr.org/2013/08/11/211130501/the-algae-is-coming-but-its-impact-is-felt-far-from-water.

Altieri, Andrew H., and Keryn B. Gedan. "Climate Change and Dead Zones." *Global Change Biology,* vol. 21, no. 4, 10 Aug. 2014, pp. 1395–1406.

Aronsohn, Marie D. "Studying Bioluminescent Blooms in the Arabian Sea." *State of the Planet,* 7 Dec. 2017, blogs.ei.columbia.edu/2017/12/04/studying-bioluminescent-blooms-arabian-sea.

Berwyn, Bob, et al. "Tiny Pink Algae May Have a Big Role in the Arctic Melting." InsideClimate News, 4 Jan. 2017, insideclimatenews.org/news/24062016/tiny-pink-algae-snow-arctic-melting-global-warming-climate-change.

Bothwell, Max L., et al. "The Didymo Story: The Role of Low Dissolved Phosphorus in the Formation of Didymosphenia Geminata Blooms." *Diatom Research,* vol. 29, no. 3, 4 Mar. 2014, pp. 229–236.

Chapra, Steven C., et al. "Climate Change Impacts on Harmful Algal Blooms in U.S. Freshwaters: A Screening-Level Assessment." *Environmental Science & Technology,* vol. 51, no. 16, 2017, pp. 8933–8943.

"Climate Change and Harmful Algal Blooms." Environmental Protection Agency, 9 Mar. 2017, www.epa.gov/nutrientpollution/climate-change-and-harmful-algal-blooms.

"Collateral Consequences: Climate Change and the Arabian Sea." Lamont-Doherty Earth Observatory, 4 Dec. 2017, www.ldeo.columbia.edu/news-events/collateral-consequences-climate-change-and-arabian-sea.

Conniff, Richard. "The Nitrogen Problem: Why Global Warming Is Making It Worse." *Yale Environment 360,* 7 Aug. 2017, e360.yale.edu/features/the-nitrogen-problem-why-global-warming-is-making-it-worse.

Danovaro, Roberto, et al. "Sunscreens Cause Coral Bleaching by Promoting Viral Infections." *Environmental Health Perspectives,* 2008.

Dell'Amore, Christine. "River Algae Known as Rock Snot Boosted by Climate Change?" *National Geographic,* 12 Mar. 2014.

Embury-Dennis, Tom. "'Dead Zone' Larger than Scotland Found by Underwater Robots in Arabian Sea." *The Independent,* 27 Apr. 2018, www.independent.co.uk/environment/dead-zone-arabian-sea-gulf-oman-underwater-robots-ocean-pollution-discovery-a8325676.html.

"Fast Facts: Hurricane Costs." Office for Coastal Management, National Oceanic and Atmospheric Administration, https://coast.noaa.gov/states/fast-facts/hurricane-costs.html.

"Forest Fertilization in British Columbia." *British Columbia,* Ministry of Forests and Range, https://www2.gov.bc.ca/assets/gov/environment/natural-resource-stewardship/land-based-investment/forests-for-tomorrow/fertilizationsynopsisfinal.pdf.

Ganey, Gerard Q., et al. "The Role of Microbes in Snowmelt and Radiative Forcing on an Alaskan Icefield." *Nature Geoscience,* vol. 10, no. 10, 2017, pp. 754–759.

"Impacts of Climate Change on the Occurrence of Harmful Algal Blooms." Environmental Protection Agency Office of Water, May 2013, www.epa.gov/sites/production/files/documents/climatehabs.pdf.

Jones, Ashley M. Environmental Protection Agency, 22 Aug. 2016, blog.epa.gov/

blog/2016/08/from-grasslands-to-forests-nitrogen-impacts-all-ecosys tems.

Miller, Melissa A., et al. "Evidence for a Novel Marine Harmful Algal Bloom: Cyanotoxin (Microcystin) Transfer from Land to Sea Otters." *PLOS One,* 10 Sept. 2010, journals.plos.org/plosone/article?id=10.1371%2Fjournal.pone.0012576.

Ogden, Nicholas H., et al. "Estimated Effects of Projected Climate Change on the Basic Reproductive Number of the Lyme Disease Vector Ixodes Scapularis." *Environmental Health Perspectives,* 14 Mar. 2014.

O'Hanlon, Larry. "The Brown Snot Taking over the World's Rivers." *BBC Earth,* 29 Sept. 2014, www.bbc.com/earth/story/20140922-green-snot-takes-over-worlds-rivers.

Paerl, Hans W., and Jef Huisman. "Blooms Like It Hot." *Science,* 4 Apr. 2008.

Pelley, Janet. "Taming Toxic Algae Blooms." American Chemical Society, *ACS Central Science, 2* (5), pp 270–273, 12 May 2016. (How nitrogen and phosphorus runoff will make algal blooms.)

Preece, Ellen. "Transfer of Microcystin from Freshwater Lakes to Puget Sound, WA, and Toxin Accumulation in Marine Mussels." *EPA Presentation Region 10 HAB Workshop,* US EPA, 29 Mar. 2016, www.epa.gov/sites/production/files/2016-03/documents/transfer-microcystin-freshwater-lakes.pdf.

Stibal, Marek, et al. "Algae Drive Enhanced Darkening of Bare Ice on the Greenland Ice Sheet." *Geophysical Research Letters,* vol. 44, no. 22, 28 Nov. 2017, pp. 11463–11471.

"Vermont Repeals Felt Sole Ban." *American Angler,* 23 June 2016, www.americanangler.com/vermont-repeals-felt-sole-ban.

"*Vibrio* Species Causing Vibriosis." Centers for Disease Control and Prevention, 19 Apr. 2018, www.cdc.gov/vibrio/index.html.

Welch, Craig. "Climate Change Pushing Tropical Diseases Toward Arctic." *National Geographic,* 14 June 2017.

Wheeler, Timothy. "2010 food poisoning cases linked to Asian bacteria in raw oysters." *Bay Journal,* 18 May 2016, www.bayjournal.com/article/2010_food_poisoning_case_linked_to_asian_bacteria_in_raw_oysters. (Data on cases of *vibrio* infection.)

Yardley, Jim. "To Save Olympic Sailing Races, China Fights Algae." *New York Times,* 1 July 2018, www.nytimes.com/2008/07/01/world/asia/01algae.html.

Yirka, Bob. "Algae Growing on Snow Found to Cause Ice Field to Melt Faster in Alaska." *Phys.org,* 19 Sept. 2017, phys.org/news/2017-09-algae-ice-field-faster-alaska.html.

Zielinski, Sarah. "Ocean Dead Zones Are Getting Worse Globally Due to Climate Change." *Smithsonian,* 10 Nov. 2014.

"*Zooplankton:* Noctiluca Scintillans." University of Tasmania, Institute for Marine and Antarctic Studies, 2 Feb. 2013, www.imas.utas.edu.au/zooplankton/image-key/noctiluca-scintillans.

CHAPTER 6: GEOENGINEERING

Arrigo, Kevin R., et al. "Melting Glaciers Stimulate Large Summer Phytoplankton Blooms in Southwest Greenland Waters." *Geophysical Research Letters*, vol. 44, no. 12, 31 May 2017, pp. 6278–6285.

Biello, David. "Controversial Spewed Iron Experiment Succeeds as Carbon Sink." *Scientific American*, 18 July 2012.

Bishop, James K. B., and Todd J. Wood. "Year-Round Observations of Carbon Biomass and Flux Variability in the Southern Ocean." *Global Biogeochemical Cycles*, vol. 23, no. 2, May 2009.

Cumming, Vivien. "Earth – How Hot Could the Earth Get?" *BBC Earth*, 30 Nov. 2015, www.bbc.com/earth/story/20151130-how-hot-could-the-earth-get.

Disparte, Dante. "If You Think Fighting Climate Change Will Be Expensive, Calculate the Cost of Letting It Happen." *Harvard Business Review*, 5 July 2017.

Dodd, Scott. "DMS: The Climate Gas You've Never Heard Of." *Oceanus Magazine*, 17 July 2008, www.whoi.edu/oceanus/feature/dms--the-climate-gas-youve-never-heard-of.

Fountain, Henry. "How Oman's Rocks Could Help Save the Planet." *Gulf News*, 26 Apr. 2018, https://gulfnews.com/news/gulf/oman/how-oman-s-rocks-could-help-save-the-planet-1.2213007.

Glennon, Robert. "The Unfolding Tragedy of Climate Change in Bangladesh." *Scientific American Blog Network*, 21 Apr. 2017, blogs.scientificamerican.com/guest-blog/the-unfolding-tragedy-of-climate-change-in-bangla desh. (Data on Bangladeshis to be displaced by sea rise.)

Gramling, Carolyn, et al. "Tiny Sea Creatures Are Making Clouds over the Southern Ocean." *Science*, 9 Dec. 2017, www.sciencemag.org/news/2015/07/tiny-sea-creatures-are-making-clouds-over-southern-ocean.

Grandey, B. S., and C. Wang. "Enhanced Marine Sulphur Emissions Offset Global Warming and Impact Rainfall." *Scientific Reports*, 21 Aug. 2015.

Johnston, Ian. "The Cost of Climate Change Has Been Revealed, and It's Horrifying." *The Independent*, 16 Nov. 2016, www.independent.co.uk/envi ronment/global-warming-climate-change-world-economy-gdp-smaller-12-trillion-a7421106.html.

Jones, Nicola. "Abrupt Sea Level Rise Looms As Increasingly Realistic Threat." *Yale Environment 360*, 5 May 2016, e360.yale.edu/features/abrupt_sea_level_rise_realistic_greenland_antarctica.

Kintisch, Eli. "Should Oceanographers Pump Iron?" *Science*, 30 Nov. 2007.

Kohnert, Katrin, et al. "Strong Geologic Methane Emissions from Discontinuous Terrestrial Permafrost in the Mackenzie Delta, Canada." *Nature News*, 19 July 2017, www.nature.com/articles/s41598-017-05783-2.

Lear, Caroline H., et al. "Breathing More Deeply: Deep Ocean Carbon Storage during the Mid-Pleistocene Climate Transition." *Geology*, 1 Dec. 2016. (On the 100,000-year cycle of algae deposition.)

Ocean Portal Team. "Ocean Acidification." Smithsonian Institution Ocean, https://ocean.si.edu/ocean-life/invertebrates/ocean-acidification. (On the rate of ocean acidification.)

Zielenski, Sarah. "Iceland Carbon Capture Project Quickly Converts Carbon Dioxide Into Stone." Smithsonian.com, 9 June 2016, www.smithsonianmag.com/science-nature/iceland-carbon-capture-project-quickly-converts-carbon-dioxide-stone-180959365/.

Epilogue

"Carbon fibers, made from algae oil." *Algae Industry Magazine,* 26 Nov. 2018, http://www.algaeindustrymagazine.com/carbon-fibers-made-from-algae-oil/. (Reporting on the word of Dr. Thomas Bruck at the Algae Cultivation Center of the Technical University of Munich in coordination with chemists at the university.)

"Climate Change—A Feast of Ideas." Australian Meat Processor Organization, 21 Nov. 2016, www.youtube.com/watch?v=X_JQJeZeizs. (*Asparagopsis* fed to cattle to reduce methane emissions.)

Couso, Inmaculada, et al. "Synergism between Inositol Polyphosphates and TOR Kinase Signaling in Nutrient Sensing, Growth Control, and Lipid Metabolism in Chlamydomonas." *The Plant Cell,* vol. 28, no. 9, 6 Sept. 2016, pp. 2026–2042. (How signaling leads to higher levels of lipid accumulation.)

Hernandez, I., et al. "Pricing of Monoclonal Antibody Therapies: Higher If Used for Cancer?" *American Journal of Managed Care,* Feb. 2018. (For the price of recombinant protein monoclonal antibodies.)

Houwat, Igor, and Taylor Weiss. "Better Together: A Bacteria Community Creates Biodegradable Plastic with Sunlight." *MSU-DOE Plant Research Laboratory,* 23 Oct. 2017, prl.natsci.msu.edu/news-events/news/better-together-a-bacteria-community-creates-biodegradable-plastic-with-sunlight.

Kennedy, Merrit. "Surf and Turf: To Reduce Gas Emissions from Cows, Scientists Look to the Ocean." NPR, 3 July 2018, www.npr.org/sections/the-salt/2018/07/03/623645396/surf-and-turf-to-reduce-gas-emissions-from-cows-scientists-look-to-the-ocean.

Kinley, Robert D., et al. "The Red Macroalgae Asparagopsis taxiformis Is a Potent Natural Antimethanogenic That Reduces Methane Production during in Vitro Fermentation with Rumen Fluid." *Animal Production Science,* vol. 56, no. 3, 9 Feb. 2016, p. 282.

Li, Xixi, et al. "*Asparagopsis Taxiformis* Decreases Enteric Methane Production from Sheep." *Animal Production Science,* vol. 58, no. 4, Aug. 2016, p. 681.

"Major Cuts of Greenhouse Gas Emissions from Livestock within Reach (Major Facts and Findings)." Food and Agricultural Organization of the United Nations, 26 Sept. 2013, www.fao.org/news/story/en/item/197608/icode.

Pancha, Imran, et al. "Target of rapamycin (TOR) signaling modulates starch accumulation via glycogenin phosphorylation status in the unicellular red alga Cyanidioschyzon merolae." *The Plant Journal,* 23 Oct. 2018.

Rasala, Beth A., and Stephen P. Mayfield. "Photosynthetic Biomanufacturing in Green Algae; Production of Recombinant Proteins for Industrial, Nutritional, and Medical Uses." *Photosynthesis Research,* vol. 123, no. 3, Mar. 2015, pp. 227–239.

Sanchez-Garcia, Laura, et al. "Recombinant Pharmaceuticals from Microbial Cells: A 2015 Update." *Microbial Cell Factories,* BMC, 9 Feb. 2016, www.ncbi.nlm.nih.gov/pmc/articles/PMC4748523.

Schwartz, Zane. "How One Researcher Is Fighting Cow Farts—and Climate Change—by Feeding the Gassy Beasts Seaweed." *National Post,* 21 Oct. 2016, nationalpost.com/news/world/how-one-researcher-is-fighting-cow-farts-and-climate-change-by-feeding-the-gassy-beasts-seaweed.

Sorigué, Damien, et al. "An algal photoenzyme converts fatty acids to hydrocarbons." *Science,* 1 Sept. 2017, vol. 357, no. 6354, pp. 903–907.

Taunt, Henry, et al. "Green Biologics: The Algal Chloroplast as a Platform for Making Biopharmaceuticals." *Bioengineered,* vol. 9, no. 1, 2018, www.tandfonline.com/doi/full/10.1080/21655979.2017.1377867.

"Turning Green Algae into Colostrum-like Protein for Infants—Triton Algae Innovations." National Science Foundation and Triton Algae Innovations, 1 May 2018, www.youtube.com/watch?v=9oOIQuWLRAs.

Yan, Na, et al. "The Potential for Microalgae as Bioreactors to Produce Pharmaceuticals." *International Journal of Molecular Sciences,* vol. 17, no. 6, 17 June 2016, p. 962.

Index